The Fisherman's Cause

Atlantic Commerce and Maritime Dimensions of the American Revolution

In the first book-length examination of the connections between the commercial fishing industry in colonial America and the American Revolution, Christopher P. Magra places the origins and progress of this formative event in a wider Atlantic context. *The Fisherman's Cause* utilizes extensive research from archives in the United States, Canada, and the U.K. in order to take this Atlantic approach. Dried, salted cod represented the most lucrative export in New England. The fishing industry connected colonial producers to transatlantic markets in the Iberian Peninsula and the West Indies. Parliament's coercive regulation of this branch of colonial maritime commerce contributed to colonists' willingness to engage in a variety of revolutionary activities. Colonists then used the sea to resist British authority. Fish merchants converted transatlantic trade routes into military supply lines, and they transformed fishing vessels into warships. Fishermen armed and manned the first American Navy, served in the first coast guard units, and fought on privateers. These maritime activities helped secure American independence.

Christopher P. Magra completed his Ph.D. at the University of Pittsburgh. He is now an Assistant Professor of Early American/Atlantic History and Director of the Atlantic History Center at California State University, Northridge. He has published articles related to maritime history in the *International Journal of Maritime History*, the *New England Quarterly*, and the *Northern Mariner*. The Canadian Nautical Research Society honored him with the Keith Matthews Award for the best 2007 article published in the *Northern Mariner*.

The Fisherman's Cause

Atlantic Commerce and Maritime Dimensions of the American Revolution

CHRISTOPHER P. MAGRA
California State University, Northridge

CAMBRIDGE UNIVERSITY PRESS
Cambridge, New York, Melbourne, Madrid, Cape Town, Singapore, São Paulo, Delhi

Cambridge University Press
32 Avenue of the Americas, New York, NY 10013-2473, USA

www.cambridge.org
Information on this title: www.cambridge.org/9780521518383

First published 2009

Printed in the United States of America

A catalog record for this publication is available from the British Library.

Library of Congress Cataloging in Publication data

Magra, Christopher Paul.
The fisherman's cause : Atlantic commerce and maritime dimensions of the American Revolution / Christopher P. Magra.
p. cm.
Includes index.
ISBN 978-0-521-51838-3 (pbk.)
1. Cod fisheries – United States – History – 18th century. 2. United States – History – Revolution, 1775–1783 – Economic aspects. 3. United States. Continental Navy. I. Title.
SH351.C5M34 2009
973.3′11–dc22 2008044609

ISBN 978-0-521-51838-3 hardback

Contents

Abbreviations

AA	Force, Peter. *American Archives*, Series 4–5, Vols. 1–9, Washington, D.C.: M. St. Clair Clarke and Peter Force, 1837–53.
EVREC	*Early Vital Records of Essex County, Massachusetts to About 1850*. Vols. 1–3. Salem: Newcomb & Gauss, 1903.
IJMH	*The International Journal of Maritime History*.
JAB	Smith, Phillip Chadwick Foster, ed. *The Journals of Ashley Bowen (1728–1813)*. Vols. 1–2, Portland, Maine: The Colonial Society of Massachusetts, 1973.
JCC	Ford, Worthington C., ed. *Journals of the Continental Congress, 1774–1789*. Vols. 1–34. Washington, D.C.: U.S. Government Printing Office, 1904–37.
JDPL	James Duncan Phillips Library, Salem, Massachusetts.
JEPCM	Lincoln, William, ed. *The Journals of Each Provincial Congress of Massachusetts in 1774 and 1775, and of the Committee of Safety, with an Appendix*. Boston: Dutton and Wentworth, 1838.
JHRM	*Journals of the House of Representatives of Massachusetts, 1762–1779*. Vols. 38–55. Boston: Massachusetts Historical Society, 1968–90.
MDHS	Marblehead Museum & Historical Society, Marblehead, Massachusetts.
MHS	Massachusetts Historical Society, Boston, Massachusetts.
MPR	*Massachusetts Probate Records Middlesex & Essex Counties, Selected Years (CD-ROM)*. Provo, UT: MyFamily.com, Inc., 2000.

MSA	Massachusetts State Archives, Boston, Massachusetts.
MSSRW	*Massachusetts Soldiers and Sailors of the Revolutionary War*, Secretary of the Commonwealth, comp. Vols. 1–17. Boston: Wright and Potter Printing Co., 1896–1908.
MUMHA	Memorial University, Maritime History Archives, St. John's, Newfoundland.
NA	National Archives (formerly the Public Records Office), Kew, England.
	ADM – Records of the Navy Board and the Board of Admiralty.
	CO – Records of the Colonial Office, Commonwealth and Foreign and Commonwealth Offices, Empire Marketing Board, and related bodies.
	SP – Records assembled by the State Paper Office, including papers of the secretaries of state up to 1782.
	T – Records created and inherited by the Treasury Board.
NDAR	Clark, William Bell, et al., eds. *Naval Documents of the American Revolution.* 11 vols. to date. Washington, D.C.: U.S. Government Printing Office, 1964–.
NEQ	*The New England Quarterly.*
PDBP	Simmons, R.C. and P.D.G. Thomas, eds. *Proceedings and Debates of the British Parliaments respecting North America, 1754–1783.* Vols. 1–5. Millwood, NY: Kraus International Publications, 1982–1986.
PGW	Abbot, W.W. and Dorothy Twohig, eds. *The Papers of George Washington: Revolutionary War Series.* Vol. 1. Charlottesville: University Press of Virginia, 1985.
WMQ	*The William & Mary Quarterly.*

Introduction

> The history of Sea Power is largely, though by no means solely, a narrative of contests between nations, of mutual rivalries, of violence frequently culminating in war. The profound influence of sea commerce upon the wealth and strength of countries was clearly seen long before the true principles which governed its growth and prosperity were detected. To secure to one's own people a disproportionate share of such benefits, every effort was made to exclude others, either by the peaceful legislative methods of monopoly or prohibitory regulations, or, when these failed, by direct violence. The clash of interests, the angry feelings roused by conflicting attempts thus to appropriate the larger share, if not the whole, of the advantages of commerce, and of distant unsettled commercial regions, led to wars.[1]

After a long transatlantic journey, which included a sailing time of approximately eight weeks, the brig *Pitt Packet* was homeward bound on April 22, 1769. A rising April sun would have melted fog and warmed chilled hands in the early spring morning as the brig's crew made preparations to enter the harbor at Marblehead, Massachusetts.[2] Thomas Power,

[1] Alfred Thayer Mahan, *The Influence of Sea Power Upon History, 1660–1783* (New York: Dover Publications, Inc., 1987; originally published by Little, Brown, and Company, Boston, in 1890), 1.

[2] Marblehead weather patterns for the month of April were observed and recorded by Ashley Bowen in his journals. *JAB*, I-II. While Bowen did not record the weather for April 1769, he did make such observations for the two previous Aprils. See, ibid., I, 153, 175. The *Pitt Packet's* crew comprised a captain, a mate, a master mariner, a cook, and four common seamen. L. Kinvin Wroth and Hiller B. Zobel, eds., *Legal Papers of John Adams*, Vol. 2 (Harvard University Press, 1965), 277, 313–320. The average sailing time was based on the round trip from Boston to London in the eighteenth century. Ian K. Steele, *The English Atlantic 1675–1740: An Exploration of Communication and Community* (Oxford University Press, 1986), 57.

the *Pitt Packet's* master, probably shouted orders for tack and sheet lines to be hauled in and secured as the brig angled toward the harbor mouth.[3] A chorus of barefoot seamen would then sing pulling songs, or sea chanties, in order to lend a cadence to barehanded tacking labors.[4] Singing and working, anticipating long-absent family and friends, the *Pitt Packet's* crew prepared to come home.

The brig belonged to Robert "King" Hooper, proprietor of one of the largest fish merchant houses in Marblehead, which was then the principal commercial fishing port in New England in terms of capital investment, number of vessels, and manpower.[5] The *Pitt Packet*'s crew had transported their dried, salted cod to overseas markets in Spain, as did a smaller percentage of other colonial crews in the eighteenth century.[6] There, the processed fish was sold to merchants in ports such as Bilbao and Cadiz. Typically, crews returned with fruits, loads of salt, lines of credit, and manufactured goods from England. On this particular voyage, the *Pitt Packet* was returning directly from Cadiz with salt for Marblehead's commercial fishing industry.[7]

The brig did not immediately reach Marblehead's harbor that chilly April morning, however. Dawn's early light also illuminated bent backs and busy hands on board the British naval frigate *Rose*, which was patrolling Atlantic waters along New England's coastline. The *Rose*'s Captain, Benjamin Caldwell, sent Lieutenant Henry Gibson Panton with several armed men to board the *Pitt Packet* early Saturday morning on the pretext of searching for contraband. But, Caldwell's real intent was to press men into naval service.[8] With its sails trimmed in preparation

[3] See, John Harland, *Seamanship in the Age of Sail: An Account of the Shiphandling of the Sailing Man-of-War, 1600–1860, Based On Contemporary Sources* (Naval Institute Press, 1984), 145–154.

[4] Frederick Pease Harlow, *Chanteying Aboard American Ships* (Meriden, CT: Barre Publishing Company, Inc., 1962).

[5] For more on Hooper, see the short biography in *JAB*, Vol. 2, 661. For more on Marblehead, see Christopher P. Magra, "'Soldiers . . . Bred to the Sea': Maritime Marblehead, Massachusetts and the Origins and Progress of the American Revolution," *New England Quarterly*, Vol. LXXVIII, No. 4 (December 2004), 531–562.

[6] A majority of colonial shipments of dried cod went to the West Indies in the eighteenth century. See chapter four.

[7] For the *Pitt Packet*'s cargo on the day of its impressment and seizure, see Customs Commissioners to Salem Customs Officers, April 27, 1769, Salem Custom House Record Book, 1763–1772, folders 280–281, JDPL.

[8] On Captain Caldwell's orders, and his intent to impress, see the eyewitnesses' testimony compiled in "Adams' Minutes of the Trial," Wroth and Zobel, eds., *Legal Papers of John Adams*, Vol. 2, 293–322.

for homecoming, the Marblehead brig could not evade the *Rose* and its press gang.

The crew-members on board the *Pitt Packet*, however, were not willing to surrender once Panton and the press gang boarded the brig. Michael Corbet, Pierce Fenning, John Ryan, and William Conner, Irishmen who called Marblehead home, picked up "fish gig, musket, hatchet, and harpoon" and stood ready to forcibly resist impressment.[9] It is likely that the brig served double duty in Hooper's employ as a fishing vessel and a trading vessel, and the work tools had simply been left on board during the trade voyage. In any case, the four common seamen, those directly in danger of being impressed, armed themselves with fishing implements and retreated inside the brig.

The resistance quickly escalated. After a tense standoff amidst piles of salt in the forepeak, Corbet drew a line on the floor using handfuls of the preservative and dared the press gang to cross it. Panton unwisely took up this challenge, stepped over the salt line, and advanced toward "the Ship's People." Corbet then launched the harpoon he had been holding, which struck Panton in the neck severing his jugular vein. The British lieutenant fell, his men carried him to the main deck, and he bled to death. During the ensuing mêlée, two Marblehead mariners were shot and severely wounded. The colonial laborers were then arrested and tried for murder.[10] Their trial gained notoriety throughout the colonies.[11] In the end, the maritime laborers' defense attorney, who was none other than John Adams, was able to get the men acquitted on the basis that Panton's death occurred as a result of justifiable homicide in self-defense.

The *Pitt Packet* affair, colonial fish merchants, and those who labored in the colonial fishing industry are not typically part of the American independence story. The British North American mainland colonies that became the United States of America have been portrayed as being fundamentally rural and agrarian, with inconsequential port cities.[12] Farmers

[9] Ibid., 277.

[10] For more on this event, see Magra, "'Soldiers . . . Bred to the Sea'," 531–562.

[11] *Boston Chronicle*, April 27, 1769; *Boston Gazette*, May 1, 1769; for coverage in the *Boston Evening Post*, the *New York Journal*, and the *Pennsylvania Chronicle*, see Oliver Morton Dickerson, comp., *Boston Under Military Rule, 1768–1769, As Revealed In A Journal of the Times* (Westport, CT: Greenwood Press, 1971; orig. pub., 1936), 94–95, 104–105.

[12] Daniel Scott Smith maintains that an expansive and accessible western frontier effectively prevented large-scale population concentrations and urban development during the colonial era. Smith, "A Malthusian-Frontier Interpretation of U.S. Demographic

and farming tend to dominate accounts of European settlement and economic development in North America.[13] Political and economic issues surrounding the ability to buy and sell borderlands between British settlement and Indian country have been shown to have influenced the decision of colonial land speculators and agriculturalists to resist British authority during the late eighteenth century imperial crisis.[14] In the words of one notable early American historian, the American Revolution should be seen "as a consequence of the forty-year-long effort to subject the Ohio Country, and with it the rest of the Transappalachian west, to imperial control."[15] Other colonial farmers remained relatively isolated from the outside world in rural communities and only reluctantly agreed to fight for American independence when the British Army was on their doorsteps and town meetings were banned.[16] Farmers then became minutemen, filled the ranks of the Continental Army under George Washington, and heroically fought against a powerful British military to defend their liberties and livelihoods in the face of tyranny.[17] In this interpretation, the Atlantic Ocean represents nothing more than a large liminal space

History Before c. 1815," in Woodrow Borah, et al. eds., *Urbanization in the Americas* (Ottawa: History Division, National Museum of Man, 1980), 15–24.

[13] According to Richard B. Sheridan, "the agricultural sector engaged 80–90 percent of the work force." As for the export sector, he writes: "Foreign trade, however important it was as an energizing force, constituted only around 9–12 percent of colonial gross output." Sheridan, "The Domestic Economy," in Jack P. Greene and J.R. Pole, eds., *Colonial British America* (The Johns Hopkins University Press, 1984), 43. For more along these lines, see Edwin J. Perkins, *The Economy of Colonial America* (New York: Columbia University Press, 1980).

[14] See, for example, Alan Taylor, *The Divided Ground: Indians, Settlers, and the Northern Borderland of the American Revolution* (New York: Alfred A. Knopf, 2006); and Woody Holton, *Forced Founders: Indians, Debtors, Slaves, and the Making of the American Revolution in Virginia* (University of North Carolina Press, 1999).

[15] Fred Anderson, *Crucible of War: The Seven Years' War and the Face of Empire in British North America, 1754–1766* (New York: Vintage Books, 2000), xx.

[16] See Richard L. Bushman, *King and People in Provincial Massachusetts* (North Carolina University Press, 1985); Robert A. Gross, *The Minutemen and Their World* (New York: Hill and Wang, 1976); and John Shy, *Toward Lexington: The Role of the British Army in the Coming of the American Revolution* (Princeton University Press, 1965).

[17] Don Higginbotham, *The War of American Independence: Military Attitudes, Policies, and Practice, 1763–1789*, 2nd ed. (Northeastern University Press, 1983). Higginbotham's account is typical of the land-centered analytical framework used by most military historians. He discusses "the provincialism," "the rural isolation, the traditions of localism," that defined "a predominantly agrarian society" in British North America. He also maintains "the War of Independence was for Americans mainly a defensive type of struggle waged on the patriots' soil." According to Higginbotham, soldiers in the Continental Army "were mainly farmers, blacksmiths, tanners, and artisans." Higginbotham, *The War of American Independence*, 7, 11, 12, 13, 21.

separating lands where history unfolded.[18] The maritime dimensions of the American Revolution fade to black.[19]

Farmers played formative roles in the origins and progress of the American Revolution, yet, the *Pitt Packet* affair serves as a stark reminder that colonial resistance to British authority during the Revolutionary Era cannot be fully explained without investigating why those who made their living from the sea participated in this resistance. Similarly, the progress of this formative event cannot be completely understood without coming to terms with how those tied to the sea contributed to the war for independence. In the case of New England, it was both an epicenter of revolutionary fervor and the headquarters of commercial fishing in colonial America.

It is widely acknowledged that New Englanders played important roles in bringing about the imperial crisis that separated the colonies from the mother country at the end of the eighteenth century. Riots directed against the sovereignty of the British government were particularly prevalent in New England throughout the 1760s and early 1770s. In these mob activities, effigies and customs vessels were burned; customs officials and royal governors were forced to watch their property being destroyed; monopolized tea became flotsam and jetsam; and British soldiers were harassed to the point at which they were willing to shoot into a crowd of unarmed colonists.[20] Moreover, the idea to boycott British manufactured goods and use consumer power as a political weapon began in Massachusetts;

[18] For a thorough discussion of this historical interpretation in general, and for arguments against it, see Bernhard Klein and Gesa Mackenthun, eds., *Sea Changes: Historicizing the Ocean* (New York: Routledge, 2003); and Martin W. Lewis and Kären E. Wigen, *The Myth of Continents: A Critique of Metageography* (University of California Press, 1997). Also, see W. Jeffrey Bolster, "Putting the Ocean in Atlantic History: Maritime Communities and Marine Ecology in the Northwest Atlantic, 1500–1800," *The American Historical Review*, Vol. 113, No. 1 (February 2008), 19–47;" and Marcus Rediker, "Toward a People's History of the Sea," in David Killingray, Margarette Lincoln, and Nigel Rigby, eds., *Maritime Empires: British Imperial Maritime Trade in the Nineteenth Century* (Rochester, NY: Boydell, in association with the National Maritime Museum, 2004), 195–206.

[19] To date, there is not a single overview of the maritime dimensions of the American Revolution.

[20] Alfred F. Young, *The Shoemaker and the Tea Party: Memory and the American Revolution* (Boston: Beacon Press, 1999); Andrew S. Walmsley, *Thomas Hutchinson and the Origins of the American Revolution* (New York University Press, 1999); Edmund S. Morgan and Helen M. Morgan, *The Stamp Act Crisis: Prologue to Revolution*, 3rd ed. (University of North Carolina Press, 1995); Gary B. Nash, *The Urban Crucible: Social Change, Political Consciousness, and the Origins of the American Revolution* (Harvard University Press, 1979); Dirk Hoerder, *Crowd Action in Revolutionary Massachusetts, 1765–1780* (New York: Academic Press, 1977); Eric Foner, *Tom Paine and*

militia units were formed here with the intention of resisting British authority at a minute's notice; elites in this region established the first committees of correspondence to unite colonists in opposition; and the "shot heard around the world" was fired here.[21]

Furthermore, men from New England were exceptionally active during the American Revolutionary War. Communities in this area supplied more manpower to the war effort on a consistent yearly basis for the duration of the conflict than any other region.[22] In any given year, no colony/state ever provided more men for service in the Continental Army than Massachusetts.[23] In addition, three of the four major generals who served under George Washington at the start of the war were from New England, and seven of the eight brigadier generals hailed from this region.[24] Also, four of the seven members of the first Naval Committee appointed by the Continental Congress in 1775 were New Englanders.[25] It is no coincidence that these revolutionaries lived and worked in a region that was the center of commercial fishing in colonial America.

The cod fishing industry, in particular, was one of the most valuable extractive industries in all of colonial America, and it was the single most lucrative export business in New England.[26] On the eve of the American

Revolutionary America (Oxford University Press, 1976); Bernard Bailyn, *The Ordeal of Thomas Hutchinson* (Harvard University Press, 1974); and Thomas C. Barrow, *Trade and Empire: The British Customs Service in Colonial America, 1660–1775* (Harvard University Press, 1967).

[21] T. H. Breen, *The Marketplace of Revolution: How Consumer Politics Shaped American Independence* (Oxford University Press, 2004); Gross, *The Minutemen and Their World*; Pauline Maier, *From Resistance to Revolution: Colonial radicals and the development of American opposition to Britain, 1765–1776* (New York: Vintage Books, 1972); and Shy, *Toward Lexington*.

[22] In 1775, New England provided 91% of the manpower for the Continental Army, while the Mid-Atlantic region provided the remainder. In 1776, New England provided 50% of the same type of manpower; 28% for the Mid-Atlantic region; and 22% for the Southern region. In 1777, the equivalent figures were 40%; 25%; and 35%. In 1778, 39%; 24%; and 37%. In 1779, 41%; 26%; and 33%. In 1780, 45%; 33%; and 22%. In 1781, 55%; 26%; and 19%. In 1782, 52%; 23%; and 25%. In 1783, 54%; 27%; and 19%. Thomas L. Purvis, *Revolutionary America, 1763–1800* (New York: Facts on File, 1995), 234–240, table 8.53. These figures are based on the number of men actually furnished, not the quotas required.

[23] Purvis, *Revolutionary America*, 234–240, table 8.53.

[24] Higginbotham, *The War of American Independence*, 89–90.

[25] Samuel Eliot Morison, *John Paul Jones, A Sailor's Biography* (Boston: Little, Brown and Company, 1959), 36.

[26] Economic historians consider commercial fishing to have been an extractive industry similar to lumbering and mining. See John J. McCusker and Russell R. Menard, *The Economy of British America, 1607–1789* (University of North Carolina Press, 1985), chapter fifteen. Entrepreneurs invested capital and hired workers in order to extract a resource from the ocean for processing and export to overseas markets.

Revolution, between 1768 and 1772, colonial merchants sold fish overseas valued at £152,155, which represented thirty-five per cent of the region's total export revenue.[27] Moreover, the cod fisheries employed a significant portion of colonial New England's population. Of the 581,100 people living in this region in 1770, an estimated 10,000 men found employment in this sector of the economy.[28] These 10,000 men represented eight per cent of the adult male working population.[29] Such levels of labor and capital impressed overseas observers such as Adam Smith, who penned the following at the start of the American Revolutionary War: "[T]he New England fishery in particular was, before the late disturbances, one of the most important, perhaps, in the world."[30]

Massachusetts was the site of the principal fishing ports and shipping centers in New England throughout the colonial period. By themselves, Marblehead and Gloucester, Massachusetts, accounted for sixty per cent of all the fish caught annually in the entire New England region.[31] Coastal communities in this colony were responsible for shipping nearly

[27] McCusker and Menard, *The Economy of British America*, 108, table 5.2. By comparison, settlers in Atlantic Canada (Quebec, Nova Scotia, and Newfoundland) exported fish, their leading commodity, worth £133,932 in 1768, or seventy-three per cent of the total export revenue for the region. Ibid., 115, table 5.4. At the same time, Middle Colonies (New York and Pennsylvania) exported grains worth £379,380, or seventy-two per cent of the total export revenue. Ibid., 199, table 9.3. The Upper South (Maryland and Virginia) exported tobacco worth £756,128, or seventy-two per cent. Ibid., 130, table 6.1. The Lower South (North Carolina, South Carolina, and Georgia) exported rice worth £305,533, or fifty-five per cent. Ibid., 174, table 8.2. The West Indies exported muscovado sugar worth £2,762,250, or seventy-one per cent. Ibid., 160, table 7.3. Colonial New England maintained a smaller, more diversified export sector than most of the other regional economies. Yet, even in this diversified sector, fish brought in fifteen per cent more annual revenue than the second most lucrative export, livestock.

[28] Robert G. Albion, William A. Baker, and Benjamin W. Labaree, *New England and the Sea* (Mystic, CT: Mystic Seaport Museum, Inc., 1972), 29–30. This figure is conservative. One contemporary estimate placed the number of workers "employ'd in the Cod fishery" as high as 13,000. *Boston Evening Post*, January 20, 1766. For the population figure, see McCusker and Menard, *The Economy of British America*, 103, table 5.1.

[29] The eight per cent calculation was made following Daniel Vickers's method of first factoring a fifty-five per cent male population and then factoring a forty per cent demographic of men aged fifteen to forty-five: "the male working population." Daniel Vickers, *Farmers and Fishermen: Two Centuries of Work in Essex County, Massachusetts, 1630–1830* (University of North Carolina Press, 1994), 194n.

[30] Adam Smith, *An Inquiry into the Nature and Causes of the Wealth of Nations*, Max Lerner, ed. (New York: The Modern Library, 1937; orig, pub. in 1776), 544–545.

[31] Albion, Baker and Labaree, *New England and the Sea*, 30. Because the Massachusetts fishing industry took such a majority of the share of the catch from the entire region, and maintained such a large proportion of fishermen and fishing vessels, Daniel Vickers asserts that during the colonial period the New England and Massachusetts fishing industries "can be treated as roughly equivalent." Vickers, *Farmers and Fishermen*, 154, table 4.

100 per cent of the total quintals (112 pounds dry weight; pronounced "kentals") of cod exported from New England to Southern Europe between 1771 and 1772. The same ports were responsible for shipping forty-five per cent of the total quintals of cod to the West Indies during this period. Combined, these coastal communities shipped eighty-five per cent of all the fish caught in the colonies in the early 1770s.[32] Although there were certainly commercial fisheries in other colonies, none were as extensive or commercially viable as those in Massachusetts.[33]

Having said this, the commercial cod fishing industry had a wide impact on New England's economic life. Colonists relied on the revenue from the Atlantic cod trade, and they exchanged the dried, salted commodity itself, in order to purchase imports. These imports included trade goods and raw materials from the West Indies and Southern Europe, foodstuffs from the Mid-Atlantic and Southern regions, and manufactured goods from Great Britain. Such imports were distributed throughout New England. Molasses from the West Indies was especially crucial in terms of fueling distilleries located in all the New England colonies. In addition, the fishing industry benefited important regional enterprises such as shipyards, lumber mills and ropewalks, and local artisans such as carpenters, blacksmiths, shipriggers, and sailmakers. Without fish, the most valuable export in all of New England, considerably fewer imports would have entered the region, and there would have been less demand for domestic goods and services.[34]

Cursory correlations have been made between the fact that the commercial fishing industry was headquartered in New England and the fact

[32] British North American Customs Papers, 1765–1774, MHS. The fact that a quintal was the equivalent of 112 pounds should be taken with a grain of salt. Units of measurement were anything but standardized in the early modern Atlantic world. See, John J. McCusker's remarks in "Roundtable Reviews of Peter E. Pope, *Fish into Wine: The Newfoundland Plantation in the Seventeenth Century* with a Response by Peter E. Pope," in *International Journal of Maritime History*, Vol. 17, No. 1 (June, 2005), 251–261.

[33] When asked "from what Ports do the Shipping employed in the New England fishery fit out," a witness called before Parliament to testify on the nature of the colonial fishing industry in 1775 replied, "the greater part of them from Marblehead, Salem, and Cape Anne [Gloucester], for the Cod Fishery." *AA*, Series 4, Vol. 1, 1644.

[34] Boston merchants observed in 1781 that "the various mechanics [i.e. urban artisans] necessarily employed in the building, rigging and fitting out such a number of vessels, must without it [i.e. the commercial fishing industry] be destitute of subsistence. And the great quantities of provisions expended by our fishermen, and the timber made use of in building the vessels, together with the staves, hoops, &c. made use of in the exportation of the fish and oil, will convince us that the loss of the Fishery must essentially affect our inland brethren." Resolve of a Boston town meeting at Faneuil Hall, December 11, 1781. *Gentlemen, the inhabitants of the town of Boston*... (Boston: Benjamin Edes and Sons, 1781), Early American Imprints, Series I: Evans #17105.

that there was exceptional Revolutionary activity in the region. For example, we know that Marblehead, Massachusetts, the foremost fishing port in the thirteen British North American colonies on the eve of the Revolutionary War, was second only to Boston in forming a committee of correspondence.[35] Mobilization rates for Marblehead were also higher than the average mobilization rates for colonial farming communities. It was typical for twenty-two to thirty-five per cent of rural towns' adult male population to participate in the Revolution.[36] Yet this key fishing community sent nearly thirty-nine per cent.[37] Moreover, Marbleheaders commonly re-enlisted for at least one additional tour of duty after their initial experience in the war.[38] In short, there has been some recognition of the absence of an ephemeral "*rage militaire*" among men from the foremost fishing community in colonial America that burned out quickly after the fiery passions of 1775–76 had cooled.[39] Having said this, a book-length academic study of the military mobilization of colonial fishing communities in the Revolutionary War has never been attempted.[40]

One might expect naval historians who specialize in the Revolution to investigate all of the maritime aspects related to this conflict. And to a certain extent they have linked the sea to military events that occurred during the war. These scholars focus on the U.S. Navy's formation, command

[35] Ronald N. Tagney, *The World Turned Upside Down: Essex County During America's Turbulent Years, 1763–1790* (West Newbury, MA: Essex County History, 1989), 68–73; and George Athan Billias, *General John Glover and His Marblehead Mariners* (New York: Henry Holt and Company, 1960), 31. Committees of correspondence were organized in the colonies to facilitate communication among disaffected elites and to organize resistance to British authority.

[36] Higginbotham, *The War of American Independence*, 389–390.

[37] William Arthur Baller, "Military mobilization during the American Revolution in Marblehead and Worcester, Massachusetts" (Ph.D. Dissertation, Clark University, 1994), 20. Baller does not systematically investigate the reasons for this high mobilization.

[38] Walter Leslie Sargent, "Answering the Call to Arms: The Social Composition of the Revolutionary Soldiers of Massachusetts, 1775–1783" (Ph.D. Dissertation, University of Minnesota, 2004), 226, table 6.5; and Baller, "Military mobilization," 27–28. Neither Sargent nor Baller examines the occupational identities of Revolutionary soldiers. Nor do they fully probe the wartime mobilization of labor and capital in the fishing industry.

[39] For more on the concept of *rage militaire*, see Charles Royster, *A Revolutionary People at War: The Continental Army and American Character, 1775–1783* (University of North Carolina Press, 1979), esp. 25–53.

[40] The "new" military history being done on the Revolution, which examines the social composition of the rank and file along with the various relationships between the Continental Army and the American society that created and perpetuated it, has yet to investigate the mobilization of maritime industries such as the cod or whale fisheries, or shipbuilding. For example, there is not a single chapter on maritime enterprise or mariners in John Resch and Walter Sargent, eds., *War and Society in the American Revolution: Mobilization and Home Fronts* (Northern Illinois University Press, 2007).

structure, and tactics.[41] They have further concentrated on military leadership in a biographical mode.[42] There has even been an inquiry into the identities of maritime prisoners of war.[43] However, this otherwise excellent body of work has little to say about the maritime origins of the conflict.[44] Neither do these experts say much about the military mobilization of the fishing industry during the Revolution.

Experts who specialize in the history of oceanic fisheries have probed the relationship between commercial fishing and conflict in the modern era. It is well documented, for example, that the clash between Iceland and Great Britain during the second half of the twentieth century originated in competing claims to fishing waters in the Atlantic Ocean.[45] Fish corporations in the UK and a variety of commercial fisheries in Iceland

[41] For example, see E. Gordon Bowen-Hassell, Dennis M. Conrad, and Mark L. Hayes, *Sea Raiders of the American Revolution: The Continental Navy in European Waters* (Washington: Naval Historical Center, Department of the Navy, 2003); Robert Gardiner, ed., *Navies and the American Revolution, 1775–1783* (London: Chatham Publishing, in association with the National Maritime Museum, 1996); Nicholas Tracy, *Navies, Deterrence, & American Independence: Britain and Seapower in the 1760s and 1770s* (University of British Columbia Press, 1988); *Maritime Dimensions of the American Revolution* (Washington, D.C.: Naval History Division, Department of the Navy, 1977) (author's note: this a thirty-six page pamphlet consisting of two short conference papers devoted to the U.S. Navy and three brief comments); William M. Fowler, Jr., *Rebels Under Sail: The American Navy During the Revolution* (New York: Charles Scribner's Sons, 1976); William Bell Clark, *George Washington's Navy: Being An Account of his Excellency's Fleet in New England* (Louisiana State University Press, 1960); Gardner Weld Allen, *A Naval History of the American Revolution*, Vols. 1–2, (Boston: Houghton Mifflin Company, 1913); and Charles Oscar Paullin, *The Navy of the American Revolution: Its Administration, Its Policy, and Its Achievements* (Chicago: The Burrows Brothers, Co., 1906).

[42] See, for example, Charles E. Claghorn, *Naval Officers of the American Revolution: A Concise Biographical Dictionary* (Metuchen, NJ: Scarecrow Press, 1988); Morison, *John Paul Jones*; William Bell Clark, *Lambert Wickes, Sea Raider and Diplomat: The Story of a Naval Captain of the American Revolution* (Yale University Press, 1932); and James L. Howard, *Seth Harding, Mariner: A Naval Picture of the Revolution* (Yale University Press, 1930).

[43] See, Francis D. Cogliano, *American Maritime Prisoners in the Revolutionary War: The Captivity of William Russell* (Annapolis, MD: Naval Institute Press, 2001).

[44] Notable exceptions include Neil R. Stout, *The Royal Navy in America, 1760–1775: A Study of Enforcement of British Colonial Policy in the Era of the American Revolution* (The United States Naval Institute, 1973); and Carl Ubbelohde, *The Vice-Admiralty Courts and the American Revolution* (University of North Carolina Press, 1960). These excellent works do not discuss commercial fishing.

[45] Hannes Jónsson, *Friends in Conflict: The Anglo-Icelandic Cod Wars and the Law of the Sea* (London: Hurst & Co., 1982); and Bruce Mitchell, "Politics, Fish, and International Resource Management: The British-Icelandic Cod War," *Geographical Review*, Vol. 66, No. 2 (April 1976), 127–138.

sought to safeguard their respective access to fishing waters by lobbying for, and mobilizing, state protection in the form of coast guard vessels (in the case of Iceland) and naval vessels (in the case of the UK). These military forces then engaged each other in a series of four conflicts from 1952 to 1976 that have been called the "Cod Wars." Military conflicts originating in the modern era over cod should not be overly surprising. Modern consumers hungrily enjoy fish, and commercial fishing has remained a lucrative business. The various oceanic fisheries around the globe produced ninety-five million tons of fish for sale in 2004, valued at $84.9 billion.[46]

Fisheries historians who specialize in colonial America have not spent much time on the connections between commercial fishing and the American Revolution. Raymond McFarland devotes only one short chapter to this formative event in his broad overview of all types of commercial fishing throughout New England's history.[47] Harold Adams Innis provides only a few pages on the conflict in an examination primarily focused on Newfoundland.[48] James G. Lydon concentrates solely on the cod trade between New England and the Iberian Peninsula.[49] Daniel Vickers offers only a few comments on the Revolution in an exceptionally detailed analysis of the social relations of work in the industry.[50] Most recently, W. Jeffrey Bolster has focused exclusively on linking commercial fishing to the depletion of oceanic life in colonial New England.[51] Yet, despite

[46] The Food and Agriculture Organization of the United Nations, "The State of World Fisheries and Aquaculture," (2006), http://www.fao.org/docrep/009/A0699e/A0699E04.htm#4.1.2.

[47] Raymond McFarland, *A History of the New England Fisheries* (New York: D. Appleton & Company, 1911).

[48] Harold Adams Innis, *The Codfisheries: The History of an International Economy* (Yale University Press, 1940).

[49] James G. Lydon, "Fish for Gold: The Massachusetts Fish Trade with Iberia, 1700–1773," *New England Quarterly*, Vol. 54, No. 4 (December 1981), 539–582; "North Shore Trade in the Early Eighteenth Century," *American Neptune*, Vol. 28 (1968), 261–274; and "Fish and Flour for Gold: Southern Europe and the Colonial American Balance of Payments," *Business History Review*, Vol. 39 (1965), 171–183.

[50] Vickers, *Farmers and Fishermen*. Elsewhere, the foremost expert on the subject has asserted his belief that "the locals" in colonial Marblehead, Massachusetts, the foremost fishing port in North America, were "cautious about political engagement" and remained "more nervous about engaging in radical action" during the Revolution. See Daniel Vickers, "*Young Men and the Sea: Yankee Seafarers in the Age of Sail*: A Roundtable Response," *International Journal of Maritime History*, Vol. 17, No. 2 (December 2005), 365.

[51] W. Jeffrey Bolster, "Putting the Ocean in Atlantic History."

this lack of attention, at least one fisheries expert has acknowledged that "international disputes over fishery matters have a rather long history."[52] Moreover, dried, salted cod represented the fourth most valuable export in the Western Hemisphere for the British Empire on the eve of the American Revolution.[53] Therefore, it should not be terribly shocking to learn that the merchants and laborers involved in this lucrative maritime business played formative roles in the birth of the United States of America.

Only a few explicitly maritime histories have made direct linkages between transatlantic commerce in general and the Revolution. "New" maritime historians have studied the social history of seafaring in order to explain the motivations behind the decisions of maritime laborers who fought in the conflict.[54] They detail the ways in which capitalist social relations and impressment raids generated discontent among multi-ethnic, or "motley," maritime workers, which contributed to transnational working-class resistance to British authority during the late eighteenth century. Additionally, academic research has been done on privateers, or commerce raiders. They have been seen as privately owned and publicly sanctioned business ventures aimed at capturing enemy shipping, usually commercial vessels, for sale to prize agents.[55] The profit motive is typically considered to be privateers' *raison d'être*. Moreover, recent research into the British blockade of the North American coast has revealed that this naval action was successful in disrupting the American economy, particularly its agricultural sector.[56] While the collective weight of this

[52] Hiroshi Kasahara, "International Fishery Disputes," in Brian J. Rothschild, ed., *World Fisheries Policy: Multidisciplinary Views* (University of Washington Press, 1972), 17.

[53] In order, sugar, tobacco, and grain were more valuable. Stephen J. Hornsby, *British Atlantic, American Frontier: Spaces of Power in Early Modern British America* (University Press of New England, 2005), 26–28, esp. figure 2.1.

[54] See Peter Linebaugh and Marcus Rediker, *The Many-Headed Hydra: Sailors, Slaves, Commoners, and the Hidden History of the Revolutionary Atlantic* (Boston: Beacon Press, 2000); and Jesse Lemisch, *Jack Tar vs. John Bull: The Role of New York's Seamen in Precipitating the Revolution* (New York: Garland Publishers, 1997). For a concise discussion of new maritime history, see Margaret S. Creighton and Lisa Norling, eds., *Iron Men, Wooden Women: Gender and Seafaring in the Atlantic World, 1700–1920* (The John Hopkins University Press, 1996), xi.

[55] Reuben Elmore Stivers, *Privateers & Volunteers: The Men and Women of Our Reserve Naval Forces, 1766 to 1866* (Annapolis: Naval Institute Press, 1975); William Bell Clark, *Ben Franklin's Privateers* (Louisiana State University Press, 1956); Louis F. Middlebrook, *Maritime Connecticut During the American Revolution, 1775–1783*, 2 vols. (Salem, MA: The Essex Institute, 1925); and Octavius Thorndike Howe, *Beverly Privateers In The American Revolution* (Cambridge, MA: John Wilson and Son, 1922).

[56] Richard Buel, Jr., *In Irons: Britain's Naval Supremacy and the American Revolutionary Economy* (Yale University Press, 1998). For a more traditional interpretation on the

scholarship helps us come to terms with the ocean-centered nature of the Revolution, none of this academic work focuses on the fishing industry.

More than 230 years have passed since the *Pitt Packet* affair and Adam Smith's remarks. To date, no one has systematically investigated the connections between commercial fishing and the American Revolution. A thorough examination of these connections adds to our understanding of the ways in which oceans have influenced the course of human history.[57] The Atlantic Ocean fundamentally affected the origins and progress of the Revolution. Moreover, this investigation situates this well-studied event in a newer, wider Atlantic context.

An Atlantic approach is utilized here in order to widen the framework for analyzing the American Revolution. Traditionally, scholars have relied on local, regional, and imperial approaches in their interpretations of the Revolution. British historians since Sir Lewis Namier have originated the late-eighteenth-century imperial crisis in internal political rivalries within Parliament itself. Any study of the origins of the Revolution, Namier and his students believed, had to start and end in Whitehall.[58] Similarly, American historians came to believe that colonial rebellion could only be explained by investigating micro-level forces within the colonies. They found that overpopulation and unemployment generated large numbers of disaffected, listless youths in colonial towns who were in search of a cause. In addition, deferential politics reigned supreme in certain areas, and townsfolk simply acquiesced to revolutionary leaders they viewed as their social betters.[59] Collectively, these studies emphasize local, micro-level factors that brought about the rupture between core and periphery.

The "imperial school" of historians, by contrast, has pushed for a broader history of the origins of the American Revolution. Scholars from George Louis Beer, Leonard Labaree, and Charles M. Andrews to Ian

blockade's ineffectiveness, see David Syrett, *The Royal Navy in American Waters, 1775–1783* (Brookfield, VT: Gower, 1989).

[57] For more on the agency of oceans, see the references in footnote 15. Also, see Daniel Vickers, "Beyond Jack Tar," *William & Mary Quarterly*, Vol. 50, No. 2 (April, 1993), 418–424; and Steele, *The English Atlantic 1675–1740*, 3–18.

[58] Sir Lewis Namier, *England in the Age of the American Revolution* (London: MacMillan and Co., Limited, 1930); and idem., *The Structure of Politics at the Accession of George III*, 2 Vols. (London: MacMillan and Co., Limited, 1929).

[59] Richard L. Bushman, *King and People in Provincial Massachusetts* (North Carolina University Press, 1985); William Pencak, *War, Politics, & Revolution in Provincial Massachusetts* (Northeastern University Press, 1981); Richard Buel, Jr., *Dear Liberty: Connecticut's Mobilization for the Revolutionary War* (Wesleyan University Press, 1980); and Gross, *The Minutemen and Their World*.

R. Christie and Benjamin W. Labaree firmly situate the thirteen North American colonies within a distinctly British political and economic system in which colonists stubbornly refused to pay their fair share of post–Seven Years' War debt, which brought down the wrath of the British government on their heads.[60] In brief, colonists' recalcitrant relationship to the multi-faceted, institutional framework of empire generated tensions that tore the empire apart.[61]

It is becoming increasingly clear, however, that the Atlantic Ocean connected colonists in North America to a wider world beyond imperial borders.[62] This "ocean-centered" insight is changing the way historians write about the Revolution. Whereas town histories have focused on local, domestic factors that influenced colonists' decisions to resist British authority, and imperial historians emphasized transatlantic causes of the Revolution that were particular to an Anglo-American context,

[60] Examples of the "imperial school" scholarship include Ian R. Christie and Benjamin W. Labaree, *Empire or Independence, 1760–1776* (New York: W. W. Norton & Company, Inc., 1976); James A. Henretta, *"Salutary Neglect": Colonial Administration Under the Duke of Newcastle* (Princeton University Press, 1972); Lawrence Henry Gibson, *The British Empire Before the American Revolution*, Vols. 1–15 (New York: Alfred A. Knopf, 1939–1970); Jack M. Sosin, *Agents and Merchants: British Colonial Policy and the Origins of the American Revolution, 1763–1775* (University of Nebraska Press, 1965); Charles M. Andrews, *The Colonial Background of the American Revolution* (Yale University Press, 1931); Leonard Labaree, *Royal Government in America: A Study of the British Colonial System Before 1783* (Yale University Press, 1930); and George Louis Beer, *The Old Colonial System, 1660–1754* (New York: The MacMillan Company, 1912). For more recent work that maintains the imperial school's focus on revolutionary causations within the British Empire, see Andrew Jackson O'Shaughnessy, *An Empire Divided: The American Revolution and the British Caribbean* (University of Pennsylvania Press, 2000); Alison Gilbert Olson, *Making the Empire Work: London and American Interest Groups, 1690–1790* (Harvard University Press, 1992); and Marc Egnal, *A Mighty Empire: The Origins of the American Revolution* (Cornell University Press, 1988).

[61] It is possible to situate Oliver M. Dickerson in this scholarly camp. He investigates imperial trade and shipping regulations and finds that they did not adversely impact colonial economic life. Therefore, he concludes that such regulations cannot be seen as the economic origins of the Revolution. Oliver M. Dickerson, *The Navigation Acts and the American Revolution* (University of Pennsylvania Press, 1954). Dickerson does not, however, consider the impact of these regulations on the colonial cod fisheries.

[62] Definitions of Atlantic history include Alison Games, "Atlantic History: Definitions, Challenges, and Opportunities," *American Historical Review*, Vol. 111, No. 3 (June 2006), 741–757; Bernard Bailyn, *Atlantic History: Concept and Contours* (Harvard University Press, 2005); David Armitage, "Three Concepts of Atlantic History," in David Armitage and Michael Braddick, eds., *The British Atlantic World, 1500–1800* (London: Palgrave, 2002), 11–27; and Nicholas Canny, "Writing Atlantic History; or, Reconfiguring the History of Colonial British America," *Journal of American History*, Vol. 86, No. 3 (December 1999), 1093–1114.

Atlantic historians focus on supra-imperial forces that pressurized the British Empire until it cracked.[63] Whereas previous scholars have used the nation-state as the fundamental framework of analysis, Atlantic academics highlight those processes that crossed political borders but remained primarily within the geographic limits of the regions intimately connected to the Atlantic Ocean.

The following study connects the British North American colonies to the wider Atlantic world and situates the origins and progress of the Revolution within this ocean-centered context. An Atlantic lens is necessary to investigate the links between commercial fishing and the Revolution, as fishermen and their catches habitually traveled in and out of the political borders of the British Empire, and so did arms and ammunition during the war. Local and imperial histories of the Revolution cannot fully capture this transatlantic movement of peoples and goods or the repercussions of this maritime commerce.

The Fisherman's Cause defends a two-fold argument pertaining to the why and how of the American Revolution. The British government's efforts to control and command colonists' commercial use of the Atlantic Ocean stimulated colonial resistance to British authority.[64] The colonists then used the ocean to achieve their independence by mobilizing maritime commercial assets for war. Simply put, the origins and progress of the American Revolution cannot be fully understood without coming to terms with its maritime dimensions.

This book is divided into three parts. Part One examines the nature of the cod fish that became such a valuable Atlantic commodity. It also investigates the group identities of colonial fish merchants and fishermen and discusses the ways in which each group contributed to the expansion of a colonial business that rivaled the mother country's own. Part Two

[63] See Eliga H. Gould and Peter S. Onuf, eds., *Empire and Nation: The American Revolution in the Atlantic World* (Johns Hopkins University Press, 2005); and Linebaugh and Rediker, *The Many-Headed Hydra*.

[64] The perceived right of the English government to control access to the Atlantic Ocean was an old conviction dating from at least the late sixteenth century and the writings of John Dee, who argued that English monarchs maintained sovereignty over the seas. This belief was further promulgated in 1635 with the publication of John Selden's classic *Mare Clausum* (Dominion of the Sea), which Charles I commissioned in order to rebut Dutch claims to *Mare Liberum* (Free Seas). The English Republic then commissioned Marchemont Nedham to reprint Selden's work in 1652 at the start of the First Anglo-Dutch War. David Armitage, *The Ideological Origins of the British Empire* (Cambridge University Press, 2000), chapter four. It is important to note here that Dee, Grotius, Selden, and Nedham all argued over fishing rights to Atlantic waters.

explains why merchants and laborers decided to join together to resist British authority during the late-eighteenth-century imperial crisis. Part Three details the ways in which this peace-time industry was mobilized for war. The concluding section reflects on the book's broader implications.

This is the first systematic examination of the connections between the commercial fishing industry in colonial America and the American Revolution. As a result, certain lines of research have been followed more closely than others here. My goal has been to explore the ocean-centered nature of the war for American independence. Subsequent scholars may pursue other goals in linking commercial fishing and the Revolution, and I hope that they do.

John Adams long ago recognized the importance of studying the maritime dimensions of the American Revolution. Having succeeded in getting the charges against the *Pitt Packet* mariners dropped in 1769, a reflective Adams later wrote that this legal victory, this resistance, and the impressment behind both, were more significant in stirring popular sentiment for the cause of independence than the trials surrounding the Boston Massacre, in which he defended the British soldiers. In Adams's words, "Panton and Corbet ought not to have been forgotten." The Founding Father added, "Preston and his soldiers ought to have been forgotten sooner."[65]

[65] Charles Francis Adams, *The Works of John Adams, Second President of the United States*, Vol. 10 (Boston: Little, Brown and Company, 1856), 210. Captain Thomas Preston was in charge of the British soldiers who fired upon the mob in the Boston Massacre.

PART ONE

THE RISE OF THE COLONIAL COD FISHERIES

Part One examines the fish, the people, and the methods responsible for the development of the commercial fishing industry in colonial New England prior to the outbreak of the Revolutionary War.

1

Fish

> In 1784 the Massachusetts House of Representatives passed a resolution: "to hang up the representation of a cod fish in the room where the House sits, as a memorial to the importance of the cod fishery to the welfare of this Commonwealth."[1]

Cured fish of different varieties were widely traded and consumed during the eighteenth century. While visiting Boston in 1744, a wealthy Maryland traveler and physician, Alexander Hamilton, feasted upon a local favorite. "I was invited to dine with Captain Irvin," Hamilton wrote, "upon salt codfish, which here is a common Saturday dinner, being elegantly dressed with a sauce of butter and eggs."[2] At Quebec in 1776, Scottish mariner John Nicol observed that local Native Americans caught salmon on the St. Lawrence River. They then smoked the fish and used it in trade. "The Indians come alongside [moored British merchant vessels] every day with them," Nicol wrote, "either smoked or fresh, which they exchange for biscuit or pork."[3] Nine years later, Nicol sailed a trade

[1] Anne Rowe Cunningham, ed., *Letters and Diary of John Rowe: Boston Merchant, 1759–1762, 1764–1779* (Boston: W.B. Clarke Co., 1903), 330. A gold-leafed, filleted cod was also emblazoned on the Hanseatic League's medieval coat of arms. C.L. Cutting, *Fish Saving: A History of Fish Processing from Ancient to Modern Times* (New York: Philosophical Society, 1956), 143, plate 9.

[2] Doctor Alexander Hamilton, *Hamilton's Itinerarium; being a narrative of a journey from Annapolis, Maryland, through Delaware, Pennsylvania, New York, New Jersey, Connecticut, Rhode Island, Massachusetts and New Hampshire, from May to September, 1744*, edited by Albert Bushnell Hart (Saint Louis, MO: W. K. Bixby, 1907), 364.

[3] Tim Flannery, ed., *The Life and Adventures of John Nicol, Mariner* (New York: Atlantic Monthly Press, 1997; orig. pub. 1821), 33. According to Daniel K. Richter, fishing was

vessel bound from London to Canton. The crew stopped along the northwest coast of North America to catch and smoke salmon of their own for the Pacific crossing.[4] Then, while on the way back to England from Canton, Nicol fished for Albacore tuna off the coast of St. Helena. Sailors "split and hung them in the rigging to dry."[5] From gentlemen to maritime laborers, and from Europeans to Native Americans, just about everyone had a use for cured fish of different types.

Perhaps no other fish was as widely consumed in the early modern Atlantic world as cod, however. A seventeenth-century French traveler who visited Newfoundland once wrote: "One can truly say that the best trade in Europe is to go [to Newfoundland] and fish cod.... you make good Spanish coin out of it and a million men live on it in France."[6] William Poyntz, a British consul stationed in Lisbon, Portugal, wrote to Britain's Board of Trade in 1718 to inform them "that the chief Branches of the British Fish Trade to Portugal consists in dry Cod from Newfoundland & New England; of which sort large quantities are expended [i.e. sold and consumed] yearly in this Kingdom, especially in this place [i.e. Lisbon] & Oporto, from whence the inland Towns are supplied by Land & Water Carriage into the borders, & even into some parts of Spain itself."[7] West Indian planters from Great Britain's foremost sugar islands, Jamaica and Barbados, testified before Parliament in 1775 that dried, salted cod was "the meat of all the Slaves in all the West Indies."[8] Catholics and slaves, two sizable Atlantic demographics, regularly devoured large quantities of cod.

This chapter examines the biological qualities of cod to explain why this fish was such a popular food item in the eighteenth century. Consumer demand for this commodity stimulated the rise of a substantial business in colonial New England. The expansion of the colonial cod fisheries eventually put the colonies at odds with business rivals and the British state, which contributed to the late-eighteenth-century imperial crisis.

"vitally important" to native diets. It contributed as much as twenty per cent of the food intake of inland communities and much more in coastal areas. *Facing East from Indian Country: A Native History of Early America* (Harvard University Press, 2001), 57.

[4] Flannery, ed., *The Life and Adventures of John Nicol*, 94–98.

[5] Ibid., 109.

[6] Quote taken from Fernand Braudel, *The Structures of Everyday Life: Civilization & Capitalism, 15th–18th Century, Volume 1*, translated by Siân Reynolds (New York: Harper & Row, Publishers, 1979), 217–218.

[7] William Poyntz to the Lords Commissioners of Trade and Plantations, dated Lisbon, September 17, 1718, NA SP 89/26/78A.

[8] *AA*, Series 4, Vol. 1, 1722–1723, and 1731–1732.

Thus, to fully appreciate the connections between commercial fishing in early America and the Revolution, and to evaluate the role the Atlantic Ocean played in this conflict, it is first necessary to come to terms with the fish itself.

Cod used to flourish in the North Sea and Atlantic waters between Norway and Massachusetts. *Gadus morhua* is a fish characterized by five fins, a protruding upper jaw on a head that takes up one-fourth of its body length, a heavy body, and a square tail.[9] It is a demersal fish that feeds on crustaceans and other smaller forms of sea-life on the bottom of the ocean, usually along continental shelves near upwellings of cold water. Cod migrate to inshore waters to spawn, as the food is more abundant there, and they generally reproduce more efficiently in warmer waters. Cod typically grow to an adult weight of thirty pounds and a length of two feet. But, cod weighing more than 200 pounds and measuring more than six feet have been caught.[10] They are big, homely, meaty fish.

The nature of cod's flesh largely accounts for its commercial significance. It contains an unusually high amount of protein. By comparison, beef contains less than half of the protein per weight as dried, salted cod.[11] Taken wet, directly from the ocean, a cod's body weight typically contains eighteen per cent protein. Dried, salted cod filets with the water weight evaporated can be composed of eighty per cent protein.[12]

The flesh is also particularly well-suited for preservation. Herring, another fish popular to consumers in the early modern period, especially in regions around the North Sea, had a fatty, oily flesh that made it quick to spoil and unreceptive to heavy salt cures. That is why most herring was pickled or smoked and shipped only short distances.[13] Cod, by contrast, was easily preserved. Its flesh was almost purely white, containing very

[9] Brian Fagan, *Fish on Friday: Feasting, Fasting and the Discovery of the New World* (New York: Basic Books, 2006), 49, 49n, 61.

[10] Ibid., 61; Peter E. Pope, *Fish into Wine: The Newfoundland Plantation in the Seventeenth Century* (University of North Carolina Press, 2004), 24; Harold Adams Innis, *The Codfisheries: The History of an International Economy* (Yale University Press, 1940), 1–6; and Raymond McFarland, *A History of the New England Fisheries* (New York: D. Appleton & Company, publishers for University of Pennsylvania, 1911), 4.

[11] Massimo Livi-Bacci, *Population and Nutrition: An Essay on European Demographic History*, translated by Tania Croft-Murray and Carl Ipsen (Cambridge University Press, 1991), 30.

[12] Fagan, *Fish on Friday*, 62; and Mark Kurlansky, *Cod: A Biography of the Fish That Changed the World* (New York, NY: Alfred A. Knopf, 1997), 34.

[13] Fagan, *Fish on Friday*, 49, 55.

little fat. The meat was also less oily than herring. Reduced fats and oils made it possible to create a cure for the flesh of cod that would preserve it for long periods.[14]

During the eighteenth century, there were two primary methods of preserving cod in commercial usage. That is, there were two primary methods for producing and distributing cured cod in bulk quantities, as opposed to curing small amounts for subsistence. These commercial methods derived in large measure from different European taste preferences and merchants' desire to meet market demand. The French and Northern Europeans in cooler climates preferred wet, or "green," or "core," cod, which was taken directly from the sea, gutted, lightly salted, and barreled. Consumers in warmer climates in Southern Europe, by contrast, preferred cod that had been filleted, salted, pressed, and dried.[15]

There were secondary commercial methods used to dry cod as well. The fish could be taken fresh from the sea, filleted, and wind-dried on "fish flakes."[16] This curing technique was mostly utilized in colder, more northerly climates where the air somewhat freeze-dried the fish. Wind-dried cod was referred to as "stockfish."[17] "Saltfish," on the other hand, was cod that had been filleted, left in salt for several days, pressed, and spread to dry on fish flakes.[18] As a result, therefore, of its biological characteristics, cod could be processed into a very valuable trade good. Stockfish and saltfish were both capable of enduring long periods in the un-refrigerated holds of pre-modern trade vessels. Both types of dried cod further enabled consumers to store food for extended periods, and modest amounts of the fish could satiate people's daily dietary needs.[19]

There were other biological factors that made cod ideal for commerce. Left untouched by human hands or natural malady, cod reproduce in

[14] Ibid., 49, 55, 62.

[15] Cutting, *Fish Saving*, 133.

[16] According to Pope, flakes were "rough wooden platforms covered with fir boughs or birch bark." Pope, *Fish into Wine*, 28.

[17] Fagan, *Fish on Friday*, 66, 95–96; Pope, *Fish into Wine*, 11; and Innis, *Codfisheries*, 11n. Cutting advises caution on the nomenclature here: "Unless there is confusion in contemporary accounts or in those of historians, 'stockfish' seems sometimes to have been salted as well as dried, and it is possible that it was used as a generic term for all fish dried . . . " Cutting, *Fish Saving*, 120. However, Brook Watson, a merchant in London with commercial ties to fish merchants in New England, testified before Parliament in 1775 that "Stock Fish" was "Cod cured by the frost." Watson elaborated, "There is not any Salt used in curing Stock Fish." *AA*, Series 4, Vol. 1, 1643.

[18] Fagan, *Fish on Friday*, 54–55; Pope, *Fish into Wine*, 11, 28; and Vickers, *Farmers and Fishermen*, 124–125.

[19] According to Fagan, properly cured and stored dried cod can last "for five years or more." Fagan, *Fish on Friday*, 66.

prolific numbers. According to Harold Adams Innis, one of the foremost fisheries historians, "a female 40 inches long will produce 3,000,000 eggs, and it has been estimated that a 52-inch fish weighing 51 pounds would produce nearly 9,000,000."[20] Such reproduction rates meant that cod appeared for hundreds of years as a renewable, or sustainable, natural resource that could be reliably extracted for production, distribution, and consumption. Indeed, as late as 1873, Alexandre Dumas penned the following words in his *Le Grande Dictionnaire de cuisine*: "It has been calculated that if no accident prevented the hatching of the eggs and each egg reached maturity, it would take only three years to fill the sea so that you could walk across the Atlantic dryshod on the backs of cod."[21] The sheer numbers of cod reduced production costs for merchants and kept the cost at purchase low for consumers.

The oil found in cod livers was also valuable for trade. There is no evidence that its medicinal properties were widely recognized before the American Revolution.[22] Its flammable and lubricating qualities were widely known, however, and it was put to a variety of commercial and industrial applications. It was used as an illuminant in oil lamps.[23] It was also used to lubricate machines, or "trains." Indeed, "train oil" became the common nomenclature for cod liver oil.[24] In addition, it aided in the currying, or dressing, of leather.[25] During processing, cod livers were set aside and eventually placed in a "train vat." The vats were open to the sun and air, which aided the livers' decomposition. As the livers rotted over the course of weeks and even months, a layer of smelly golden brown oil rose to the surface of the vats. The oil was then skimmed off and barreled for distribution.[26]

Dried cod began to be widely consumed in Europe in the thirteenth century.[27] Stockfish was imported from Iceland, while saltfish was brought from Norway. By the early fifteenth century, cod had replaced herring, which had been sold commercially as early as the eleventh century, as the most popular fish to eat.[28] Indeed, when King Henry VIII's

[20] Innis, *Codfisheries*, 3. Fagan puts the number of eggs at five million, and he notes that it typically takes seven years for small fry to fully mature. Fagan, *Fish on Friday*, 62.
[21] Quoted in Kurlansky, *Cod*, 32.
[22] Cutting, *Fish Saving*, 152.
[23] Ibid., 151; and Fagan, *Fish on Friday*, 65.
[24] Pope, Fish into Wine, 28–29 and Innis, *Codfisheries*, 32, footnote 9.
[25] Cutting, *Fish Saving*, 152; and William B. Weeden, *Economic and Social History of New England, 1620–1789*, Vol. 2 (New York: Hillary House Publishers, Ltd., 1963), 751.
[26] Cutting, *Fish Saving*, 152.
[27] Fagan, *Fish on Friday*, 146.
[28] Ibid., 99, 103, 164.

warship *Mary Rose* sank in 1545, nearly all of the fish stowed on board to feed the soldiers and sailors was dried, salted cod, not herring.[29]

Religious norms further contributed to the popular consumption of fish in general, and, because of its biological and commercial qualities, cod in particular. By the eleventh century, Catholic dietary restrictions had made the consumption of meat a mortal sin during Lent, Advent, and various saints' days and holy days. Together, these periods of religiously enforced abstinence could comprise as much as 182 days each year.[30] The Catholic church thereby reinforced the logic of fish consumption in general and cod consumption in particular.

Cod's comparative price advantage over other sources of protein and its long shelf-life also contributed to its wide circulation among poorer white European workers and black African slaves who lived and worked around the Atlantic world. In 1762, Salem merchants paid £3.0.0 and £2.4.0 per barrel for South Carolina pork and beef, respectively.[31] At the same time, New England merchantable and refuse-grade, dried cod sold at £0.13.1 and £0.7.2 per quintal.[32] Given that a quintal equaled 112 pounds, and assuming a barrel of pork or beef equaled a British hundredweight, also 112 pounds, then pork sold at 6.4 pence per pound; beef sold at 4.7 pence per pound; merchantable, dried cod sold at 1.4 pence per pound; and refuse, dried cod sold at 0.77 pence per pound. Pork was thus 4.6 times more expensive than merchantable, dried cod and 8.3 times more costly than refuse, dried cod, while the price of beef was 3.4 times higher than merchantable, dried cod and 6.1 times higher than refuse, dried cod.[33]

In sum, cod was widely consumed in the eighteenth-century Atlantic world. Its flesh was high in protein and low in fats and oils, which allowed

[29] Ibid., 225.

[30] Ibid., 144–148, 156. Regina Grafe puts the upper limit of fast days in Spain "during the entire early modern period" at 130. Grafe, "Popish habits, nutritional need and market integration: Fasting and fish consumption in Iberia in the early modern period" (unpublished paper presented at the UCLA VonGremp Workshop in Economic and Entrepreneurial History, March 12, 2008), 15.

[31] Schooner *Esther*, 1760–68, box 5, folder 3, Timothy Orne Shipping Papers, JDPL.

[32] Daniel Vickers, "'A Knowen and Staple Commoditie': Codfish Prices in Essex County, Massachusetts, 1640–1775," *Essex Institute, Historical Collections*, Vol. CXXIV (Salem, MA: Newcomb and Gauss, 1988), 202, table 1.

[33] Grafe observes "that around the mid 17th century cod became a real substitute for beef" in Spain. She further calculates that pre-1750 Spanish consumers saw a one-to-one ratio of the cost of purchasing protein in the form of beef vs. the cost of buying protein from dried, salted cod. But, after 1750, "cod based protein was definitely cheaper." Grafe, "Popish habits," 27, 30–31, esp. figure 8.

a long-lasting cure to be used on the fish. Such preservation made cod ideal for distributors and consumers in an age without refrigeration. Prolific reproduction rates, comparatively low cost, valuable liver oil, and religious norms also made this fish a particularly popular food item.

Consumer demand for cured fish may have stimulated the production and distribution of dried, salted cod in colonial America. Such demand by itself, however, did not build a substantial maritime industry in the colonies. The following chapters will demonstrate that merchants and maritime laborers worked very hard over the course of the colonial period to make a living from the Atlantic Ocean. These efforts explain why colonists were so defensive about the commercial fishing industry during the late-eighteenth-century imperial crisis.

2

Fish Merchants

> I have heard of a certain Merchant in the west of England, who after many great losses, [was] walking upon the Sea-bank in a calm Sun-shining day; observing the smoothness of the Sea, coming in with a checkered or dimpled wave: 'Ah (quoth he) thou flattering Element, many a time hast thou enticed me to throw myself and my fortunes into thy Arms; but thou hast hitherto proved treacherous; thinking to find thee a Mother of increase, I have found thee to be the Mother of mischief and wickedness; yea the Father of prodigies; therefore, being now secure, I will trust thee no more.' But mark this man's resolution awhile after, *periculum maris spes lucri superat* [hope for money overcomes the danger of the sea].[1]

The Gerry family in Marblehead, Massachusetts, owned and operated one of the most successful fish exporting businesses in eighteenth-century New England. Thomas Gerry (1702–1774), the patriarch, was born in Newton Abbot in Devonshire, located in England's West Country, a region with historic ties to the Atlantic cod trade. He came to colonial Massachusetts as a twenty-eight-year-old master of a merchant ship in 1730 and decided to settle in Marblehead, then an up-and-coming fishing port. Four years later, he married Elizabeth Greenleaf, the daughter of a Marblehead saddler and a relation of John Elbridge, who had been collector of customs in Bristol, England, before he died and left the Greenleaf family an estate in excess of one million pounds sterling.[2]

[1] John Josselyn, English travel writer, c. 1663. Paul J. Lindholdt, ed., *John Josselyn, Colonial Traveler: A Critical Edition of Two Voyages to New-England* (University Press of New England, 1988), 27.

[2] George Athan Billias, *Elbridge Gerry: Founding Father and Republican Statesman* (New York: McGraw Hill Book Co., 1976).

Thomas quickly built a profitable import/export business in Marblehead. Elizabeth had many children, and those who survived became directly involved in the family's business.[3] The Gerry house established transatlantic trade networks with merchants in Iberian and West Indian ports in addition to England and the North American mainland colonies. Butler and Mathews of Cadiz, Edward Broome of Lisbon, Joseph Gardoqui and Sons of Bilbao, Nicholas Lane of London, and Joseph Howell of Philadelphia were several of the "citizens of the world" with whom the Gerrys routinely corresponded and did business.[4] The Gerrys exported Marblehead's best grade of dried, salted cod – the "merchantable" grade – to Catholic markets in Lisbon, Bilbao, and Cadiz to be exchanged for salt, wine, raisins, lemons, solid specie, and lines of credit.[5] The lesser grade of dried, salted cod – the "refuse" grade – was shipped to slave plantations in the West Indies for rum, sugar, and molasses.[6] Some West Indian and Iberian products were sold in the middle and southern mainland colonies for flour, grain, and other provisions. From the 1730s to the 1770s, the Gerrys' vessels, *Makepeace*, *Pretty Betsey*, *Success*, and *Rockingham*, regularly sailed these Atlantic trade routes. Moreover, the Gerrys owned substantial real estate in Marblehead, including their own wharf and a general store that sold retail trade goods imported from all over the

[3] Of eleven children, only five survived infancy. Gerry Family Geneological Records, MHS. In a letter of attorney dated April 7, 1775, the shareholders in the family business are listed as the five surviving children. Elbridge Gerry Papers to 1780, MHS.

[4] In addition, Gerry received information from the captain of one of his Iberian trade ships, Captain Alex Ross; the captain of one of his West Indian trade ships, Captain Joseph Northey; and "Masters Lynch, Killikelly and Moroney" in an unspecified location. Gerry also received information from his family members on their travels to various American ports. Letter Book of Samuel Russell Gerry, 1769–1784, MDHS, item 1913.295; and Elbridge Gerry Papers to 1780, MHS. David Hancock applied the phrase "citizens of the world" to eighteenth-century merchants. David Hancock, *Citizens of the World: London Merchants and the Integration of the British Atlantic Community, 1735–1785* (Cambridge University Press, 1995).

[5] John Josselyn, who visited the New England coast on two occasions during the seventeenth century, defined merchantable-grade, dried, salted cod c. 1667 as "being sound, full grown fish and well made up, which is known when it is clear like a Lanthorn horn and without spots." Lindholdt, ed., *John Josselyn, Colonial Traveler*, 144.

[6] Edward Payne, a Boston fish merchant, defined refuse-grade, dried, salted cod as "that being only such as is over Salted, Sun Burned, and broken, & thereby rendered unfit for any Market in Europe." "In the Preamble to a late Act of Parliament," 1764, attributed to Edward Payne, Ezekiel Price Papers, 1754–1785, MHS. One member of Parliament defined it in 1763 as "bad cod fish which will be consumed by negroes only." *PDBP*, Vol. 1, 491. Nearly 100 years earlier, Josselyn defined it as "salt burned, spotted, rotten, and carelessly ordered." Lindholdt, ed., *John Josselyn, Colonial Traveler*, 144.

Atlantic world. At his death, Thomas left an estate that included six vessels, a wharf, and a warehouse.[7] The wealth from the cod trade enabled Thomas's third son, Elbridge, to attend Harvard, become a member of the Continental Congress and signer of the Declaration of Independence, and eventually Vice President of the United States of America.[8]

This chapter focuses on maritime business matters. There are two main objectives here. The chapter establishes the group identity of fish merchants like the Gerrys, men who helped build the most valuable extractive industry and export producer in all of colonial New England. It also establishes the methods by which these sea-focused entrepreneurs made their business profitable.

The group identity of colonial fish merchants was closely linked to a New England economy transitioning toward capitalism.[9] Although a capitalist socio-economic system had emerged in England prior to the migration of the Pilgrims and the Puritans, it took time for entrepreneurs in New England to surmount the hurdles associated with scarce labor and capital.[10] Over the course of the 1700s, economic hallmarks of capitalism such as labor markets, capital markets, and commodity markets were established in New England.[11]

[7] *JAB*, II, 655. Also, see Log Book Schooner *Rockingham*, Ship's Logs box 1, 1926.99. Loan, MDHS.

[8] Billias, *Elbridge Gerry*.

[9] Academics fiercely debate the nature of capitalism and the date of its emergence in New England. For an overview, see Naomi R. Lamoreaux, "Rethinking the Transition to Capitalism in the Early American Northeast," *Journal of American History*, Vol. 90, No. 2 (September 2003): 437–461.

[10] For more on these hurdles and the ways in which colonists surmounted them, see Daniel Vickers, *Farmers and Fishermen: Two Centuries of Work in Essex County, Massachusetts, 1630–1830* (University of North Carolina Press, 1994); and Winifred Barr Rothenberg, *From Market Places to a Market Economy: The Transformation of Rural Massachusetts, 1750–1850* (University of Chicago Press, 1992). Both Vickers and Rothenberg maintain that capitalism emerged in England prior to the settlement of New England. For more on this point, see Immanuel Wallerstein, *The Modern World-System I: Capitalist Agriculture and the Origins of the European World-Economy in the Sixteenth Century* (New York: Academic Press, 1974).

[11] Scholars who associate capitalism with certain cultural values – rather than the emergence of labor markets, capital markets, and commodity markets – do not agree that the eighteenth century was a formative era. Phyllis Hunter, Stephen Innes, and John Frederick Martin, for example, maintain that Puritan religious ideals inspired entrepreneurial quests for profits from the first settlement in the 1600s. Phyllis Whitman Hunter, *Purchasing Identity in the Atlantic World: Massachusetts Merchants, 1670–1780* (Cornell University Press, 2001); Stephen Innes, *Creating the Commonwealth: The Economic Culture of Puritan New England* (New York: W.W. Norton, 1995); and John Frederick

As a result of this transitioning process, colonial fishing communities became increasingly stratified over the course of the seventeenth and eighteenth centuries. Wealth in these communities concentrated in the hands of a few, and the divide between rich and poor widened. Christine Heyrman has studied Gloucester and Marblehead, Massachusetts, the most productive commercial fishing ports in mainland North America, and she has determined that the richest ten per cent of these communities controlled an average of thirty-eight per cent of the real and personal wealth in 1690. The same decile group commanded an average of fifty-seven per cent of the same wealth in 1770.[12] Colonial fish merchants were among this privileged group. Such a wealth trend was common throughout the North American colonies prior to the American Revolution.[13]

The means of production in the colonial cod fisheries also became concentrated in the hands of a few merchants over the course of the eighteenth century. Whereas forty per cent of the fishermen in Daniel Vickers's study owned vessels before 1676, only two per cent had control of this productive asset on the eve of the American Revolution.[14] Due to certain advances in ship design, which will be discussed at greater length in the following chapter, most commercial fishing vessels had simply become too expensive for the average fisherman.

Fish merchants' conspicuous consumption further separated them from workers in the commercial fishing industry. The "codfish aristocracy" were merchants whose wealth and opulent colonial lifestyle stemmed from the Atlantic cod trade. Men such as James Bowdoin, the Faneuils (Andrew, Peter, and Benjamin Jr.), Thomas Hancock, John Jeffries, Stephen Minot, Timothy Orne, Anthony Stoddard, and Jacob

Martin, *Profits in the Wilderness: Entrepreneurship and the Founding of New England Towns in the Seventeenth Century* (University of North Carolina Press, 1991). James Henretta argues that there is no evidence of capitalist *mentalités* among farmers in New England at any point during the entire colonial period. James A. Henretta, "Families and Farms: *Mentalité* in Pre-industrial America," *WMQ*, Vol. 35, No. 1 (January 1978), 3–32. Also, see James Henretta, *The Origins of American Capitalism: Collected Essays* (Northeastern University Press, 1991). Naomi Lamoreaux sides with Henretta. She compares Bostonian merchants to Henretta's farmers and finds that colonial merchants cannot be considered truly profit motivated. Lamoreaux, "Rethinking the Transition to Capitalism in the Early American Northeast."

[12] Christine Leigh Heyrman, *Commerce and Culture: The Maritime Communities of Colonial Massachusetts, 1690–1750* (New York: W.W. Norton & Company, 1984), 415, Table I.

[13] Alice Hanson Jones, *Wealth of a Nation To Be: The American Colonies on the Eve of the Revolution* (Columbia University Press, 1980), esp. 183.

[14] Vickers, *Farmers and Fishermen*, 161.

Wendell purchased the latest fashions and refined home furnishings with profits gained in large part from the cod trade. Manicured gardens surrounded brick mansions that contained large veneered and gold-gilt clocks, mirrors, fine China, Asian quilts, fanciful curtains, paintings, and chests holding wigs and silk clothes.[15] Jeremiah Lee, one of the wealthiest fish exporters in colonial America, built an enormous mansion house in Marblehead in 1768. At a cost of £10,000, it was considered one of the most expensive homes in all of the colonies.[16] The accumulation of material possessions that exceeded needs was part-and-parcel with merchant bids for elevated social status and at least the appearance of gentility.[17]

In addition to this conspicuous consumption, the social relations of work in the colonial cod fisheries changed in ways that set fish merchant employers apart from their workers. Fish merchants were more litigious in the eighteenth century than they had been in the previous era. In the seventeenth century, when labor was scarce, merchant employers offered credit on easy terms to attract workers, and debts were rarely collected from employees through legal means. During the 1700s, in a more robust labor market, fish merchant employers became less willing to extend long-term and interest-free credit to their workers. They also began to frequently sue workers who bought goods on credit and then defaulted.[18] Courts routinely found in favor of fish merchants. These verdicts and the initial litigation sharply divided creditors from debtors. Moreover, during the seventeenth century, and well into the eighteenth century, the colonial cod fisheries operated primarily through personal, face-to-face agreements between fish merchants and fishermen.[19] However, these employers petitioned the Massachusetts government to make the social

[15] Hunter, *Purchasing Identity in the Atlantic World*, chapter four.

[16] Stephen J. Hornsby, *British Atlantic, American Frontier: Spaces of Power in Early Modern British America* (University Press of New England, 2005), 86–87; and Thomas Amory Lee, "The Lee Family of Marblehead," *Essex Institute Historical Collections*, Vol. LII (1916), 331.

[17] For more on the social and political implications associated with colonial consumer habits, see T. H. Breen, *The Marketplace of Revolution: How Consumer Politics Shaped American Independence* (Oxford University Press, 2004); and Hunter, *Purchasing Identity in the Atlantic World*.

[18] Vickers, *Farmers and Fishermen*, 153–167; and Heyrman, *Commerce and Culture*, 250.

[19] In 1768, the foremost fish merchants in Marblehead acknowledged this fact in a petition to the royal governor of Massachusetts. The petition stated that heretofore "Fishermen ship themselves on board Fishing Vessels for such time & on such Terms as is verbally agreed on." "Petition of the Selectmen of Marblehead [To Governor Francis Bernard and

relations of work contractual and, therefore, legally binding in 1768.[20] No longer were fishermen perceived as worthy of trust, and no more were work arrangements made through manly handshakes. By the Revolution, a gulf divided fishing communities and set fish merchants apart from fishermen.

Colonial merchants' business practices changed between the seventeenth and eighteenth centuries. These changes in the business of commercial fishing directly influenced colonial political relations with the mother country. During the earlier era, colonial entrepreneurs were largely dependent on West Country merchants' capital, commercial connections to overseas markets, and shipping. Small-scale fish merchants in colonial outports commonly employed coasters and fishermen to transport catches to Boston, where middlemen filled orders for, and loaded the ships of, merchants living in England's West Country. Hogsheads of dried, salted cod were then shipped overseas to foreign markets.[21] West Country control did not remain static or constant, however.

By the eighteenth century, colonial merchants had gained greater control over the production and distribution of their dried, salted cod. The English Civil War first weakened the West Country's economic control over New England. The war disrupted shipping out of England and reduced the amount of dried cod West Country merchants could supply to European markets.[22] The West Country's weakened position meant that European consumers became hungry for new suppliers. This weakened position also unleashed a scramble for cod trade profits in New England that would last for over 100 years.

the Massachusetts General Court] asking for the passage of an act relating to the Cod-fishery, with a copy of said act," 1768, MSA Collections, Vol. 66, *Maritime, 1759–1775*, 406–409.

[20] In place of any verbal agreement, which the fish merchants believed led to "many Evils," they designed a formal "agreement." "Petition of the Selectmen of Marblehead [To Governor Francis Bernard and the Massachusetts General Court] asking for the passage of an act relating to the Cod-fishery, with a copy of said act," 1768, MSA Collections, Vol. 66, *Maritime, 1759–1775*, 406–409. The codification of labor agreements between merchant and laborer in this document are very similar to Parliament's 1729 "Act for the Better Regulation and Government of the Merchants Service." For a discussion of the 1729 Act, see Marcus Rediker, *Between the Devil and the Deep Blue Sea* (Cambridge University Press, 1987), 121.

[21] Daniel Vickers, *Young Men and the Sea: Yankee Seafarers in the Age of Sail* (Yale University Press, 2005), 34–35, 42–43; and Heyrman, *Commerce and Culture*, 227–330.

[22] See the discussion in chapter four.

New England merchants lacked the shipping, capital, and overseas connections in the 1640s that were immediately necessary to fill this economic vacuum. However, London merchants were eager to break the West Country's stranglehold on the Atlantic cod trade, and they had all of the commercial necessities. These merchants began shipping manufactured goods to Boston on the first leg of what would become a regular triangular trade. In Boston, manufactured goods were exchanged with wholesalers for pre-arranged loads of dried cod, which were then shipped to ports in Iberia or the Portuguese Atlantic islands. The transatlantic shipments of cod primarily involved merchantable grades for domestic European consumption. Some refuse-grade cod was also sold in Iberia, presumably for re-export to slave plantations given the small European slave populations. Once the cod had been exchanged for fruits, solid specie, or lines of credit, the final leg of the triangle traversed to London.[23]

Colonial merchants then began building their own vessels, negotiating their own transatlantic deals, and shipping their own cod to overseas markets. During the first half of the eighteenth century, smaller fishing outports started to eliminate London middlemen based in Boston.[24] Entrepreneurs in these outports wanted greater profits from the Atlantic cod trade. In 1744, Boston selectmen bemoaned this turn of events in town meetings. They complained "both Fish and Supply is Confined to the Fishing towns who generally Send it abroad in their own Vessels, Especially Marblehead, Salem, and Plymouth, which has rendered them much Abler and Us much less to Support Our Usual proportion of the Province Tax." Ten years later, Bostonians further lamented that "the very men in Boston who heretofore supply'd the Fishery and Traders at Marblehead and elsewhere, now buy great part of their Supply of English goods from those, to whom, but a very few Years since they used to Furnish all the English or European Goods those persons had."[25] Robert Hooper and Jeremiah Lee, two of the wealthiest fish exporters who lived in Marblehead, the foremost colonial fishing community, proudly informed Lieutenant Governor Thomas Hutchinson in 1764: "We are not only largely concerned in the owning of Vessels, & employing the People

[23] Vickers, *Farmers and Fishermen*, 101–103; John J. McCusker and Russell R. Menard, *The Economy of British America, 1607–1789* (University of North Carolina Press, 1985), 99–101; Bernard Bailyn, *The New England Merchants in the Seventeenth Century*, 3rd ed. (Harvard University Press, 1995), 78–82.

[24] Vickers, *Young Men and the Sea*, 70–71; and Heyrman, *Commerce and Culture*, 330–365.

[25] *JAB*, I, 138–139.

in the Fishery, but annually ship large Quantities both to Europe and the West Indies on our own Accounts."[26] In short, while tobacco and rice planters in the southern mainland colonies continued to rely on British bottoms to freight their staples to European markets, New England fish merchants had engrossed the trade increasingly unto themselves and were shipping, in Hooper and Lee's words, "on our own Accounts."[27] Such colonial economic development explains why Thomas Gerry left the West Country and settled in Marblehead at this time.[28]

Alterations in the structure of the Atlantic cod trade meant that colonial fish merchants secured more of the profits from the trade. Profits from the sale of catches in European markets did not have to be divided between producers and shippers. Transportation costs declined as fish merchants stopped paying freight rates to overseas shippers. Information costs also decreased as middle men were eliminated. Increased business profitability did not come without additional costs and new risks, however.

The cost of doing business in the commercial cod fisheries changed over the course of the colonial period. For example, start-up and operating costs were low in the colonial fishing industry during the seventeenth century. In 1659, a shallop cost £100 in Massachusetts.[29] Such small vessels were strictly used for localized fishing operations close to shore, and they did not require much in the way of regular maintenance. Moreover, seventeenth-century workers typically purchased their own vessels. Merchants outfitted the men and the vessel and provided profit shares in

[26] Hooper Fisheries Statement, 10 January 1764, MHS, Misc. Bd. Manuscripts, 1761–1765. Lee owned, at one time or another, forty-five ships, brigs, snows, sloops, and schooners. At his death, he left an estate valued at £24,583.18.10. *JAB*, II, 665. For his part, Hooper owned a mansion, ropewalk, warehouse, and wharf, in addition to many vessels. Upon his death in 1790, Hooper's estate, valued at £5,486.6.4 was declared insolvent due to debt owed to London financiers Champion & Dickason amounting to £39,650.6.2. Ibid., 661.

[27] For the tobacco planters' reliance on British shipping well into the eighteenth century, see Jacob M. Price, *Tobacco in Atlantic Trade: The Chesapeake, London and Glasgow, 1675–1775* (Brookfield, VT: Variorum, 1995). For more on rice planters' similar dependency, see R.C. Nash, "South Carolina and the Atlantic Economy in the Late Seventeenth and Eighteenth Centuries," in Susan Socolow, ed., *The Atlantic Staple Trade*, Vol. 2 (Aldershot, England: Variorum, 1996), 353–378.

[28] It is no coincidence that other prominent fish merchant families such as the Lees and Glovers moved into Marblehead around this time as well.

[29] Daniel Vickers, "Work and Life on the Fishing Periphery of Essex County, Massachusetts, 1630–1675," in David D. Hall and David Grayson Allen, eds., *Seventeenth-Century New England* (University of Virginia Press, 1984), 98.

exchange for catches.[30] Additionally, small crews required only modest capital outlays for provisions and equipment. From 1658–1672, the cost for an outfitter such as George Corwin of Salem, Massachusetts, to outfit a shallop and a four-man crew (three fishermen and one shoreman) annually ran between £100 and £150.[31]

During the eighteenth century, by contrast, colonial fish merchants typically owned the vessels and outfitted the crews. To produce more dried, salted cod for export, eighteenth-century colonial fish merchants began using larger, more expensive vessels and employing more workers than was done in the seventeenth century. In 1775, Salem fish merchant Stephen Higginson testified before Parliament that a total of "seven hundred Vessels" were employed in the New England cod fishing industry. Of this number, "five hundred of them" averaged "from forty to seventy tons; the other two hundred from about fifteen to forty."[32] In 1763, Hooper and Lee carefully calculated the costs involved in sending a fishing vessel and crew out to sea. These successful fish merchants were eminently qualified to make such a calculation.

Hooper and Lee's calculation demonstrates that the cost of doing business in the fishing industry had increased considerably in the eighteenth century. They reckoned "the first Cost of a Banker," a vessel capable of working offshore fishing banks, most likely the larger vessels Higginson referenced, was £500, which was nearly five times the cost of a typical seventeenth-century vessel. In addition, the total capital outlay for outfitting a crew of eight for one year was £635. This figure included £160 for "Vessels ware & tare"; £195 for provisions used at sea; £40 for "Fishermen's Clothing"; and £48 for "Fishermen's provisions on shore at £6 per head." Also, in recognition of the extent to which fishermen were involved in a debt peonage relationship with merchants, Hooper and Lee factored into the costs of sending a fishing crew to sea the sum of £192 for provisions and clothing sold to the fishermen's families while they were at sea. This last figure was based on the assumption that at least half of the eight-man offshore fishing crew, most likely the older sharesmen, had "4 in his Family besides himself." In short, wealthy fish merchants viewed fishermen's families as sunk costs that needed to be factored into the total capital outlay necessary for extracting cod from offshore banks.

[30] Vickers has labeled these social relations of production "clientage." Vickers, *Farmers and Fishermen*, 153.

[31] Vickers, "Work and Life on the Fishing Periphery of Essex County," 95, footnote 1.

[32] *AA*, Series 4, Vol. 1, 1645.

These families bought goods on credit against their male relations' future earnings. Thus, a merchant had to be prepared and able to supply these items for an extended period without receiving payment. In total, Hooper and Lee calculated the cost for purchasing a single banker and outfitting a fishing crew and their family for one year to be £1,135.[33] Producing dried, salted cod for export in the eighteenth century, therefore, required a greater initial investment than it had a century earlier.

There were also greater maintenance costs associated with the production of dried, salted cod in the eighteenth century. Wooden sailing vessels required constant upkeep: leaks needed patching; new sails needed to be sown; worn planks had to be replaced; cracked spars needed fixing; scratched or dulled paint needed to be redone; worn ropes had to be replaced; ironwork needed maintenance; and broken pulleys had to be repaired. Merchants routinely paid carpenters, blacksmiths, sailmakers, painters, and other nautical artisans for their labors. Salem fish merchant Timothy Orne compensated local sailmaker Eleazar Moses £33 for the "making & mending" of the schooner *Esther*'s sails on April 15, 1763. On another occasion, August 12, 1765, Orne paid Moses £30 for "making a New Suit of Sails." One year later, the Salem fish merchant compensated local carpenter Josiah Cabot £3 for work on his schooner. At this time, Orne also paid a local blockmaker named King £2 for materials; Samuel Lascombe, a Salem joiner, £7 for his "part of Joiner's work on the Schooner *Esther*"; and Jonathan Mansfield, a Salem iron worker, the sum of £41 for general repairs to the schooner.[34] Because fish merchants

[33] "Calculations respecting outfits of a Fishing vessel," December 1763, Ezekiel Price Papers, 1754–1785, MHS. In 1755, £1,135 sterling was equivalent to $131,528.34 in year 2000 U.S. dollars. John J. McCusker, *How Much Is That In Real Money? A Historical Commodity Price Index for Use as a Deflator of Money Values in the Economy of the United States*, 2nd ed. (Worcester, MA: American Antiquarian Society, 2001), 35–36. Vickers writes that fishermen's clothing typically included "heavy boots, woolen outerclothing, and thick canvas or leather aprons called barvels." Vickers, *Farmers and Fishermen*, 122. According to Samuel Eliot Morison, "the New England fisherman's costume, until about 1830, when oilskins were adopted, was a sheep – or goat-skin jacket, and 'barvel' (leather apron), baggy calfskin trousers, yellow cowhide 'churn boots,' and tarred canvas hat, shaped like the modern sou'wester." Samuel E. Morison, *Maritime History of Massachusetts, 1783–1860*, 3rd ed. (Boston: Houghton Mifflin, 1961), 137, footnote 2. Peter E. Pope adds that their clothing was usually made from "kersey," a "coarse, narrow, woven wool cloth, usually ribbed." In addition, fishermen typically wore "a hauling hand," which was "a glove covering the palm, with the fingers protruding, used in handling fishing lines." Peter E. Pope, *Fish into Wine: The Newfoundland Plantation in the Seventeenth Century* (University of North Carolina Press, 2004), 363, table 25.

[34] Timothy Orne Shipping Papers, Schooner, *Esther*, box 5, folder 3, 1760–68, JDPL.

had become shippers as well, such repairs must have appeared as endless costs.

Merchants also faced greater risks with the production of dried cod in the eighteenth century. Most seventeenth-century fish merchants passed the risks of vessel ownership onto the fishermen. The few merchants who owned vessels at this time paid relatively little for the small craft, and risked little, as the vessels rarely left sight of shore. During the later period, by contrast, fish merchants typically owned the vessels, which were used on deep-sea fishing expeditions. Such expeditions entailed longer periods at sea and greater distances from safe harbors, which put vessels at a heightened risk of storm damage. For example, on his way home to Cape Ann from offshore fishing banks in 1769, John Lovett, skipper of the schooner *Volant*, "saw a Marblehead fisherman and Spoke with him and he told us that there Was thirteen or fourteen Sail of their fishermen Lost [as a result of a storm] as they thought."[35]

Eighteenth-century colonial fish merchants further assumed the risks associated with transatlantic shipping. There were no guarantees when it came to distributing goods across the Atlantic Ocean during the early modern Age of Sail. Vessels lacked diesel engines, depth sounders, and satellite navigation systems that could get them out of trouble. Obvious technological weaknesses further limited communication between a ship in distress and potential rescuers. Additionally, wars routinely disrupted Atlantic commerce, as rival naval vessels and privateers patrolled the usual shipping lanes, and naval convoys could interminably delay shipments. Market conditions could also change radically between the time a shipment departed and the moment the cargo reached its final port-of-call.[36] Fish merchants assumed these risks in the eighteenth century, while their predecessors had not. There is no doubt that entrepreneurs ventured and risked much to own and operate a commercial fishing business in the eighteenth century. Without their willingness to take these risks, the colonial fishing industry would have remained subservient members on the economic periphery of the Atlantic world. Colonial merchants did take these risks, however. Hope for money overcame the dangers of the sea.

[35] Log of the schooner *Volant*, 1769, George Stevens Logbooks, 1768–1774, MHS. For more on the types of severe weather fishermen encountered on deep-sea expeditions, see the discussion in chapter three.

[36] See Peter Mathias, "Risk, Credit and Kinship In Early Modern Enterprise," in John J. McCusker and Kenneth Morgan, eds., *The Early Modern Atlantic Economy* (Cambridge University Press, 2000), 15–35; and Ralph Davis, *The Rise of the English Shipping Industry in the Seventeenth and Eighteenth Centuries* (London: Macmillan & Company Limited, 1962).

Despite the additional costs of doing business in the eighteenth century, producing and selling salted cod remained a profitable enterprise. On the debit side of the balance sheet, operating costs, or "disbursements," in the colonial fishing industry included vessel wear and tear, equipment, provisions, clothing for workers, and goods sold to the workers' family members. Hooper and Lee determined that operating Marblehead's fleet of seventy-two large bankers and eighteen smaller vessels represented a total cost to colonial fish merchants of £51,318. On the credit side, Hooper and Lee further calculated that a single banker annually produced 500 quintals of merchantable-grade, dried, salted cod, which had a market value of seventeen shillings/quintal, for a total of £340. Each banker also processed 750 quintals of refuse-grade, dried, salted cod, valued at £0.12.8 per quintal, for a total of £475. A single banker was also capable of processing twenty barrels of cod liver oil, or "train oil," with a market value of forty shillings each, for a total of £40. Each banker, therefore, produced trade goods valued at £940. At the same time, each small vessel produced 400 quintals of merchantable cod worth £340 at market; 150 quintals of refuse cod valued at £95; and twelve barrels of train oil worth £24. Thus, each of these small vessels yielded commodities worth £459. In aggregate, then, Marblehead's fishing fleet of seventy-two bankers and eighteen small vessels produced goods worth £75,942 in the Atlantic marketplace each year.[37] This figure represented a net profit margin of forty-eight per cent. Such profits would have been extraordinary by most eighteenth-century business standards.[38] These lofty margins may have been the product of political propaganda, however.[39]

In 1776, Massachusetts's fish merchants calculated the "Loss On Income" that resulted from the first year of the Revolutionary War.

[37] "Calculations respecting outfits of a Fishing vessel," December 1763, Ezekiel Price Papers, 1754–1785, MHS.

[38] According to David Hancock, the standard profit margin for eighteenth-century British merchants engaged in the Atlantic slave trade was six per cent. Hancock, *Citizens of the World*, 424. Kenneth Morgan puts the typical profit margin of slavers at "10 per cent (or thereabouts)." Kenneth Morgan, *Bristol & the Atlantic Trade in the Eighteenth Century* (Cambridge University Press, 1993), 137.

[39] Hooper and Lee itemized each of the expenses colonial fish merchants incurred in their calculations. However, they did not include the "first Cost," or the start-up cost, of the bankers and the smaller vessels in their final calculation of the total cost of the fishing business in Marblehead. Moreover, the fish merchants compiled this data to pass on to Massachusetts's colonial agent to better lobby Parliament against passage of the Sugar Act. They hoped to show that the colonial fishing industry was very profitable and that the act would severely cut into their profit margin. Thus, their estimate of total cost and business profitability must be taken with a grain of salt.

Business leaders factored into their calculations the profits they lost on each fishing vessel. In Cape Ann, for example, which included three of the most commercially viable fishing ports (Gloucester, Manchester, and Beverly), and four ports of lesser consequence (Chebacco, Annisquam, Sandy Bay, and Pigeon Cove), merchants calculated that each of their "80 Sail fishing Vessels" brought in an annual net profit of £100 per vessel. Thus, the fishing fleet earned Cape Ann £8,000 each year.[40] At the same time, fish merchants in nearby Salem calculated an equivalent profit of £100 on each of their fishing schooners. With "50 Sail," these merchants figured the fleet brought into Salem £5,000 each year.[41] Assuming the Salem and Cape Anne merchants based their profit calculations on bankers, and not smaller vessels, and assuming the same profits for the Marblehead fish merchants, then the latter's seventy-two bankers brought £7,200 in net profits into Marblehead on an annual basis. Factoring £1,135 for the total cost to a fish merchant of a banker, vessel price plus operating costs, the seventy-two bankers cost Marblehead fish merchants £81,720 annually. Adding the net profit of £7,200 to this cost yields a gross profit of £88,920. By this calculation, entrepreneurs in New England's most commercially viable fishing port realized a nine per cent profit margin from the fishing business. Whether profit margins ran nine or forty-eight per cent, it is clear that the business remained profitable in the eighteenth century despite increased costs.

Colonial fish merchants adopted several new strategies to ensure that their business remained profitable despite the structural changes that occurred over the course of the colonial period. They reduced some of the total costs associated with the eighteenth-century fishing business by using the same vessels for fishing and shipping. While early fishing shallops were not suitable for transoceanic trade voyages, eighteenth-century schooners typically served a dual purpose of fishing and trading. Fish exporters owned their own vessels and paid freight rates to other vessel owners, but the former practice was more common during the eighteenth century. During the winter months, the vessels used in the Atlantic cod trade primarily freighted refuse-grade, dried, salted cod to West Indian plantations to trade for sugar products and to plantations in the Southern mainland colonies in exchange for provisions of various sorts. The

[40] "Estimate of the Loss on Income & the Trade of Cape Ann from April 1775 to April 1776," *NDAR*, Vol. 4, 1323–24.

[41] "Estimate of the Loss on Income and Trade for the Town of Salem," April 30, 1776, *NDAR*, Vol. 4, 1324–1325.

dual-usage of vessels was also a means by which profit maximizing fish merchants might avoid leaving their most capital-intensive investments, their schooners, idle and rotting in harbors.[42]

A pattern of vessels being used in this dual capacity can be found throughout eighteenth-century Massachusetts's fishing ports.[43] In Ipswich, between 1768 and 1770, the schooner *Neptune* set sail on at least five fishing expeditions, or "fares," in the spring, summer, and fall seasons. During the winter months, the same vessel left Ipswich and freighted refuse-grade, dried, salted cod to Dominica in the West Indies in addition to various coastal destinations in Virginia and Maryland. The Ipswich schooner *Polly* was taken out on six fishing fares from 1772 to 1774. Over the same period, the schooner re-exported West Indian trade goods and freighted, dried cod on trade voyages to various ports in Virginia, Maryland, and the Carolinas.[44] In Gloucester, on Cape Anne, between 1762 and 1774, fish merchant Daniel Rogers's vessels worked deep-sea fishing waters before sailing on trade voyages to Virginia, the Carolinas, and unspecified destinations in the West Indies. These Gloucester vessels included the schooners *Liberty*, *Hannah*, *George*, *Lucky*, *Rachel*, *Fame*, and *Two Brothers*.[45] In Beverly, fish merchant Thomas Davis utilized his schooners in this dual capacity during the early 1770s. His schooners *Volant*, *Swallow*, and *Swan* were taken out on fishing fares during the spring, summer, and fall, before sailing on trade voyages to Virginia each and every December.[46] In nearby Marblehead, fish merchant Richard

[42] Rhode Island merchants involved in whaling similarly used their vessels for whaling and trading. See, Virginia Bever Platt, "Tar, Staves and New England Rum: The Trade of Aaron Lopez of Newport, Rhode Island, with Colonial North Carolina," *North Carolina Historical Review*, Vol. 48 (January 1971), 1–22. I am grateful to Carl Swanson for this reference.

[43] According to Pope, seventeenth-century English fishing vessels involved in the migratory fishery at Newfoundland were also typically utilized in both fishing and trading capacities. Pope, *Fish into Wine*, 104. Barry Cunliffe has similarly noted the propensity of medieval Northern Europeans in using their fishing vessels to transport goods to overseas markets "in slack periods." Barry Cunliffe, *Facing the Ocean: The Atlantic and Its Peoples, 8000 BC–AD 1500* (Oxford University Press, 2001), 542. Fagan has found that from the fifteenth century, the Portuguese utilized the famous caravel as "warships, trading vessels, and fishing boats." Fagan, *Fish on Friday*, 208.

[44] Joshua Burnham Papers, 1758–1817, Schooner *Neptune*, [1766–1770], box 1, folder 3, JDPL; and Joshua Burnham Papers, 1758–1817, Schooner *Polly*, 1771–1776, box 1, folder 4, JDPL. For similar patterns of dual-usage in Ipswich, see Joshua Burnham Papers, 1758-1817, Schooner *Abigail*, [c. 1762], box 1, folder 1, JDPL; and Joshua Burnham Papers, 1758–1817, Schooner *Dolphin*, [c. 1762–64], box 1, folder 1, JDPL.

[45] Daniel Rogers Account Book, 1770–1790, JDPL.

[46] Thomas Davis Account Book, 1771–78, JDPL.

Pedrick's schooner *Molly* left in the winter of 1768 and 1769 with a load of refuse-grade, dried cod bound for Barbados, one of the foremost sugar islands in the British West Indies. The vessel also went on at least eleven fishing fares between 1766 and 1771.[47]

In Salem, one of the most prominent commercial ports in colonial Massachusetts, the dual usage of vessels was common.[48] Following work on fishing fares, fish merchant Timothy Orne's schooners *Esther* and *Molly* sailed on trade voyages to Maryland, the Carolinas, and Monte Christo in the West Indies throughout the 1750s and 1760s.[49] Similarly, Salem fish merchant and distillery owner Richard Derby utilized his schooners *Three Sisters* and *Three Brothers* on fishing fares to offshore banks in the 1750s, and on trade voyages to Virginia and South Carolina.[50] The dual-usage of these vessels made merchants' most capital-intensive assets valuable and productive year-round. It also meant continual wear-and-tear and maintenance fees. But merchants did not have to purchase and maintain different vessels for specialized tasks, which would have greatly increased start-up and operating costs associated with the fishing industry.

Fish merchants further reduced their operating expenses by passing some costs onto workers. Two-eighths of the profits from each and every catch was specifically allocated to cover the wear and tear on the vessel.[51] Moreover, in most eighteenth-century arrangements, verbal or otherwise, the outfitting merchant agreed to provide a vessel as well as "great" and "small" general goods for an expedition. Typically, the great general goods consisted of work-related necessities for the trip: barrels of mackerel or clams for bait, barrels of water, hogsheads of salt, candles, hooks, lines, leads, extra sail cloth, powder, shot, gloves, and mittens. The small general goods were for the most part personal items that made the trip less arduous: soap, brooms, brushes, cords of wood, paper, cider, rum, molasses, pepper, beans, flour, beef, oil, and lamps.[52] At the end of each

[47] Richard Pedrick Account Book, 1767–1784, MDHS; and Richard Pedrick Papers, schooner *Molly* [c. 1766–71], box 2, folder 28, MDHS.

[48] For more on Salem's importance as a commercial port, see Vickers, *Young Men and the Sea*.

[49] Timothy Orne Shipping Papers: schooner *Esther*, 1759–61, box 5, folder 2; schooner *Esther*, 1760–68, box 5, folder 3, JDPL; schooner *Molly*, 1751–57, box 7, folder 9; schooner *Molly*, 1751–57, box 7, folder 10; schooner *Molly*, 1758–60, box 7, folder 11; and schooner *Molly*, 1761–66, box 7, folder 12, JDPL.

[50] Richard Derby Ledger, 1757–1776, JDPL.

[51] This was common in all of the fish merchants' ledgers. See, for example, Miles Ward Ledgers, 1765–1772, JDPL.

[52] The most detailed accounts for these items include the following: Thomas Davis Account Book, 1771–78, JDPL; William Knight Account Book, 1767–1781, JDPL; and Miles Ward Ledgers, 1765–1772, JDPL.

fare, the company returned the remaining great and small general goods. The fishermen then reimbursed the merchant for any small general goods used on the fare, saving merchants the expense.

Colonial fish merchants also adopted new strategies for managing the greater risks associated with doing business in the eighteenth century. One of the ways fish merchants attempted to minimize risk was through maritime insurance, which was first introduced into colonial America in the 1720s.[53] On November 7, 1759, Cape Cod fish merchant Benjamin Bangs recorded in his diary, "I wrote for insurance on my schooner." Four days earlier, he had observed, "[I]t rains violently and this night began a most violent remarkable storm wind from SE to south. Blows a hurricane so that I believe there are but few sleepers." The next day, he noted, "[W]e see and hear of such destruction made last night by the storm that was never known in these parts." When combined with a lack of intelligence regarding the whereabouts of his schooner, the storm prompted Bangs to apply for insurance. Those same factors motivated underwriters to deny his application, however: "[S]he was reckoned to be lost and could not ensure her."[54] Similarly, Salem fish merchant Timothy Orne paid local maritime insurance agent John Higginson, Esquire, undisclosed "premiums for underwriting" between 1762 and 1764. Orne also paid premiums to Thomas King in Marblehead.[55] The Salem fish merchant then took out insurance on his schooner *Molly* for a late fishing expedition to the Grand Bank in August 1767.[56] Marblehead fish merchants Joseph Swett Jr. and Robert Hooper Jr. paid undisclosed premiums to insure their schooner *Swallow* for £250, their schooner *Phoenix* for £200, and their schooner *Salem* for £450. All of these vessels were shipping refuse cod to the West Indies.[57] Colonial fish merchants, therefore, insured their vessels whether they were operating on fishing expeditions or trading voyages, and for a small premium these entrepreneurs could protect their capital investments against total loss.

[53] For more on maritime insurance in the eighteenth century, see Christopher Kingston, "Marine Insurance in Britain and America, 1720–1844: A Comparative Institutional Analysis," *Journal of Economic History*, Vol. 67, No. 2 (May 2007): 379–409; Eric Wertheimer, *Underwriting: The Poetics of Insurance in America, 1722–1872* (Stanford University Press, 2006); Frank C. Spooner, *Risks at Sea: Amsterdam Insurance and Maritime Europe, 1766–1780* (Cambridge University Press, 1983); and John G. Clark, "Marine Insurance in Eighteenth-Century La Rochelle," *French Historical Studies*, Vol. 10, No. 4 (Autumn 1978), 572–598.

[54] Benjamin Bangs Diary, 1742–1765, MHS.

[55] Timothy Orne Shipping Papers, JDPL.

[56] Timothy Orne Shipping Papers, JDPL.

[57] Joseph Swett Jr. and Robert Hooper Jr. fonds, 1740–1747, MUMHA.

The premiums underwriters charged fish merchants for maritime insurance varied. On April 22, 1747, Bangs noted that merchants in the region "can insure none [vessels] without [a] large premium."[58] But premiums in Massachusetts seem to have leveled off at £6 during the early 1770s.[59] Variables such as climate change and war bore on underwriters' decisions. Given Hooper and Lee's estimate of £500 for a banker, a £6 premium represented just around one per cent of the cost of a deep-sea fishing vessel.

Anglo-American merchants also typically divided shares of vessel ownership to mitigate risks associated with maritime commerce during the early modern period.[60] Expenses were thereby reduced for each partner. A single investor did not have to purchase a vessel outright that might sink or be captured by pirates. Though this was not a new strategy, the increased first costs associated with schooners made partnerships that much more prudent, and eighteenth-century colonial fish merchants commonly reduced their risks in this manner. Gloucester fish merchant Ezekiel Woodward Jr. owned three-fourths of the schooner *Jolly Robbin* in 1759. Woodward's skipper/master Paul Dolliver owned the remaining quarter share of the vessel.[61] Salem fish merchant Timothy Orne owned half of the schooner *Esther* and quarter of the schooner *Sally* in 1765. Local fish merchant Benjamin Osgood owned the remainder of these vessels. Four years earlier, Orne had shared one-eighth of *Sally* with his skipper/master John Cloutman Sr. By 1762, Cloutman owned a full half of the vessel.[62] Dividing the cost of a vessel into shares meant, of course, dividing any profits made through the use of the vessel. However, having one or two partners was fiscally sensible in an era in which fishing and shipping came with great risks.

Other strategies colonial fish merchants employed to mitigate the dangers associated with sea-borne enterprise in the Age of Sail included using more than one fishing company and more than one fishing vessel. Entrepreneurs who employed more than one fishing company reduced

[58] Benjamin Bangs Diary, 1742–1765, MHS.

[59] In a rare instance in which a colonial fish merchant listed the actual premiums he paid for maritime insurance, Beverly fish merchant Thomas Davis paid the rate of £6 in premiums in 1771, 1772, and then again in 1773 for insurance on his schooners *Swallow* and *Volant* for winter trade voyages to Virginia. Thomas Davis Account Book, 1771–78, JDPL.

[60] Davis, *The Rise of the English Shipping Industry in the Seventeenth and Eighteenth Centuries*, 159, 174. For more on eighteenth-century merchant partnerships, see Hancock, *Citizens of the World*, 104–114.

[61] Aaron Parsons Sr., Shipping Papers, MSS 57, box 8, folder 7, JDPL.

[62] Timothy Orne Ledgers, 1762–1767, JDPL.

their risks by increasing the odds that fish would be caught and returned to port for processing and export. Eighteenth-century colonial fish merchants did not typically search towns to recruit fishing crews. The skipper acted as the chief labor recruiter in the fishing industry.[63] Once a skipper rounded up a "company," or fishing crew, he contracted with an outfitting merchant for employment. This contractual arrangement was either verbal or written. In 1768, Marblehead's selectmen maintained "that Fishermen ship themselves on board Fishing Vessels for such time & on such Terms as is verbally agreed on."[64] Yet, Gloucester merchant Ezekiel Woodward Jr. put into writing on September 18, 1762, to "Engage to Give Joshua Burnham Six Pounds Thirteen Shillings and 4 [pence] for Each Thousand of Cod fish said Burnham Shall Catch on Board my Schooner [*Abigail*] this fall fare on the Banks & Bring home."[65] Also, in 1758 the schooner *Molly's* crew signed a written contract with Salem merchant Timothy Orne in which the men agreed to go "upon a Fishing Voyage in Said Schooner for the Next Season upon Common Share or upon Sell Fish."[66] By these contractual means, fishermen agreed to use a fish merchant's vessel and equipment to catch fish for that merchant, and only that merchant.

Colonial fish merchants usually made such arrangements with several fishing companies at once. Beverly fish merchant Thomas Davis employed Richard Ober and Company and John Lovett and Company throughout the early 1770s.[67] Marblehead fish merchant William Knight was one of the rare vessel owners to employ his family members. Robert Knight and Company and John Chin Knight and Company both worked for the Marblehead entrepreneur during the early 1770s.[68] Knight's neighbor, fish merchant Richard Pedrick, employed John Dixey and Company along with William Brown and Company in 1770.[69] Gloucester fish merchant

[63] More will be said about labor recruitment in chapter three.

[64] "Petition of the Selectmen of Marblehead [To Governor Francis Bernard and the Massachusetts General Court] asking for the passage of an act relating to the Cod-fishery, with a copy of said act," 1768, MSA Collections, Vol. 66, *Maritime, 1759–1775*, 406–409.

[65] Joshua Burnham Papers, 1758–1817, Schooner *Abigail*, [c. 1762], box 1, folder 1, JDPL. The use of the collective plural "Banks" here, as opposed to a specific regional modifier, underscores the skipper's role in choosing the location for fishing.

[66] Timothy Orne Shipping Papers, schooner *Molly*, 1758–60, box 7, folder 11, JDPL. The share system and the count system, or "Sell Fish," will be explained later in this chapter.

[67] Thomas Davis Account Book, 1771–78, JDPL.

[68] William Knight Account Book, 1767–1781, JDPL.

[69] Richard Pedrick Account Book, 1767–1784, MDHS.

John Stevens employed Isaac Ball and Company and William Babson and Company in 1773.[70]

The companies fish merchants employed could change over time, most likely due to poor or unsatisfactory production rates. For example, Gloucester fish merchant Daniel Rogers employed Lemuel Stanwood and Company, William Foard and Company, Abraham Riggs and Company, David Walles and Company, and James Yinn and Company in 1770. The following year, Rogers retained Stanwood and Walles; he also hired John Morgan and Company, John Parsons and Company, and Alford Davis and Company.[71]

Each skipper and company worked on different vessels for their fish merchant employers. Davis owned – and employed men on – the schooners *Swallow*, *Swan*, and *Volant* during the early 1770s.[72] Knight similarly owned and operated the schooners *Barnett* and *Molly* at this time.[73] There were the schooners *Molly* and *Betsy* for Pedrick in 1770.[74] That year alone, Rogers owned, and employed men on, the schooners *Fame*, *George*, *Hannah*, *Liberty*, and *Luke*.[75] And Rogers's neighbor, Stevens, similarly owned and operated the schooners *Britannia* and *Victory* in 1773.[76] Simultaneously using several vessels reduced the risk to entrepreneurs' capital investment by ensuring a greater probability that some fish would be caught, and some fish would make it safely back to shore for processing.

Colonial fish merchants also used a pay system that helped mitigate the risks associated with the fishing industry. This system tied workers' earnings directly to Atlantic market prices of dried, salted cod.[77] Once the fish merchant knew how much and what type of fish he had to sell, as well as the current market price of each grade, a total value was placed on the catch. At that point, and only at this juncture, the crew and the

70 John Stevens Account Book, 1769–75.
71 Daniel Rogers Account Book, 1770–1790, JDPL.
72 Thomas Davis Account Book, 1771–78, JDPL.
73 William Knight Account Book, 1767–1781, JDPL.
74 Richard Pedrick Account Book, 1767–1784, MDHS.
75 Daniel Rogers Account Book, 1770–1790, JDPL.
76 John Stevens Account Book, 1769–75.
77 Fishermen in England had operated on a share system since the fifteenth century. See Maryanne Kowaleski, "The Western Fisheries," in "Fishing and Fisheries in the Middle Ages," in David J. Starkey, Chris Reid, and Neil Ashcroft, eds., *England's Sea Fisheries: The Commercial Sea Fisheries of England and Wales Since 1300* (London: Chatham Publishing, 2000), 28.

merchant sat down to come to a financial settlement. If dried cod was in great demand in overseas markets (which was known principally through correspondence between merchants and through the reports of returning ship captains) and it could fetch a high price, the value of catches and workers' profit shares were adjusted accordingly.[78] Spring fare, merchantable-grade, dried cod typically commanded the best prices on an annual basis due principally to the reputation for its quality.[79] In each annual settlement involving a share system, the fishing company's portion of the value of the fish caught on each fare typically amounted to five-eighths.[80] This portion was then divided equally among the sharesmen in the fishing company, regardless of how many fish each man actually caught. Their equivalent fish counts represented the vernacular calibration of rank, age, and experience.[81] Whereas waged labor would have meant constant operating costs regardless of market vigor, the share system ensured that merchants' profit margins remained stable. This system thereby reduced the risks and managed the costs typically associated with larger fishing crews of the eighteenth century.

Of course, the share system was not the only system colonial fish merchants used. At times, fishing crews went to sea on the count system.[82] In this less customary, more capitalistic mode of labor discipline, workers were given pre-arranged pay rates based on the number of fish caught. Usually, men agreed on a rate per every thousand fish caught. For example, Gloucester merchant John Stevens paid John Perry £19.12.10 for 3,272 fish caught on a single fare at £6 per thousand fish in 1771.[83] And, as mentioned above, Gloucester merchant Ezekiel Woodward Jr. agreed in 1762 to pay Joshua Burnham and Company £6.13.4 per every thousand fish caught."[84] Such a system encouraged an ethic of aggressive

[78] Daniel Vickers, "The Price of Fish: A Price Index for Cod, 1505–1892," *Acadiensis*, Vol. 25, No. 2 (Spring 1996), 196.

[79] Ibid., 62–81.

[80] Vickers, *Farmers and Fishermen*, 161.

[81] Vernacular calibrations can be found throughout colonial fish merchants' ledgers. For example, there were five sharesmen and only two cuttails on board, the schooner *Esther*, Benjamin Henderson, skipper, during a fall fare in 1760. The reported fish count, as opposed to the actual fish count, was that the skipper and one other man caught exactly 2,419 cod; the remaining three sharesmen each caught precisely 2,418; and the cuttails caught 1,860 and 1,524 each. Timothy Orne Ledgers, 1762–1767, JDPL.

[82] Vickers, *Farmers and Fishermen*, 161–162.

[83] John Stevens Account Book, 1769–75, JDPL.

[84] Joshua Burnham Papers, 1758–1817, Schooner *Abigail*, [c. 1762], box 1, folder 1, JDPL.

competition among fishermen and can be seen as a form of piece-work waged labor, as workers effectively engaged in a wage agreement with merchants. Colonial fish merchants may have used this system to stimulate production rates.

Only rarely did eighteenth-century fish merchants resort to "straight" wage labor.[85] Straight wages offered workers some protection from failed voyages and more predictable income. They appealed to those merchants in populous areas who were able to hire workers cheaply and thereby increase their profit margins. There is evidence that John Stevens attempted to introduce straight wages among his Gloucester workers in the second half of the eighteenth century.[86] Yet, for the most part, Atlantic cod fishermen had an aversion to straight wage labor.[87] Such antipathy was common among workers in other early modern industries.[88] Once accepted, daily wage rates tended to standardize pay in similar enterprises in proximity to one another. This meant pay by position, rather than pay by skill. In order of usage, then, colonial fish merchants employed the share system, the count system, and straight wages.[89] The share system was the most prevalent because it helped entrepreneurs maintain profit margins and manage risk.

Colonial fish merchants could resort to unscrupulous business practices to lessen business risks as well. The British consul in Lisbon reported

[85] A "straight" wage system in the commercial fishing industry involved a flat payment for labor that disregarded the number of fish caught.

[86] John Stevens paid Gideon Carter "wages" for a trip "to the Banks with Phillip Babson" in 1770. David Walles Jr. deposited in an account with Stevens his "Wages to the Banks" with John Derry Skipper at an unidentifiable date. See John Stevens Account Book, 1769–75, JDPL.

[87] Pope, *Fish into Wine*, 185, 188–189. Vickers writes that during the eighteenth century "free market relations" between labor and capital in the cod fisheries had "supplanted" more traditional relations. Vickers, *Farmers and Fishermen*, 203. But he is careful not to define these "free market relations" as specifically waged labor, nor does he dispute that fishermen preferred the share system above all else. For bold, yet unsubstantiated, claims that the Newfoundland cod fishing industry converted completely to wage labor by the mid-eighteenth century, see Innis, *The Codfisheries*, 151–152; and Ralph Greenlee Lounsbury, *The British Fishery at Newfoundland, 1634–1763* (Hamden, CT: Archon Books, 1969), 90.

[88] Christopher Hill, "Pottage for Freeborn Englishmen: Attitudes to Wage-labour," in Hill, *Change and Continuity in Seventeenth-Century England* (London: Weidenfeld and Nicolson, 1974), 219–238; and A. W. Coats, "Changing Attitudes to Labour in the Mid-Eighteenth Century," *Economic History Review*, Vol. 11, No. 1 (April,1958), 35–51.

[89] This point represents a departure from McFarland and is more in line with Vickers. See Raymond McFarland, *A History of the New England Fisheries* (New York: D. Appleton & Company, 1911), 96–97; and Vickers, *Farmers and Fishermen*, 162, esp. footnote 31.

in 1718 that shady colonial entrepreneurs, "endeavoring to be first at Market," under-cured their fish and sold it as fully cured. Other dodgy merchants "either out of self interest, or defect in their Salters, fill their fish so full of salt to make it weigh heavy." Thus, a quintal of fish could be comprised largely of salt. Others "press it [the fish] with Weights, that the fish may look fair to the eye & mouth." Then there was "the great abuse of packing new fish at each end of their hogsheads, & old in the middle."[90] Defrauding consumers enabled dishonest merchants to manage risk and maintain profits.

A certain amount of luck also helped eighteenth-century merchants manage risk.[91] This was truly the case in the fishing industry. Capricious weather patterns on the North Atlantic could suddenly destroy entire fishing fleets.[92] In such cases, the fortunate entrepreneur who chose not to invest capital in an offshore fishing expedition, or the lucky fish merchant whose skipper decided to prolong departure, could avoid financial loss. Moreover, a skipper and his crew could get lucky and stumble upon rich fishing waters, decreasing their time at sea and increasing their merchant employer's chances of timing the market.

As a group, then, eighteenth-century colonial fish merchants were wealthy, profit-oriented, risk-taking, conspicuous consumers who exhibited a spirit of individualism by concentrating economic power unto themselves. These profit-maximizing, risk-reducing entrepreneurs changed the way they did business to secure their bottom lines in the face of expanding expenses and rising risks. They decided to increase the range of their fishing operations to include deep-sea banks, which necessitated larger vessels and crews being sent to sea for longer periods of time. Colonial merchants also increasingly decided to take on the distribution responsibilities associated with the Atlantic cod trade.

Greater risks and costs were associated with these offshore expeditions and these transatlantic trade voyages. The hope for money overcame the dangers of the sea, however. Fish merchants such as Thomas Gerry invested capital, cut costs, and developed risk-managing strategies to develop a colonial industry that was profitable to individual business owners. Such

[90] William Poyntz to the Lords Commissioners of Trade and Plantations, dated Lisbon, September 17, 1718, NA SP 89/26/78A.

[91] Mathias, "Risk, Credit and Kinship in Early Modern Enterprise," 16.

[92] See, for example, the Reverend William Whitwell's sermon "Discourse Occasioned by the loss of a Number of Vessels with their Mariners Belonging to the Town of Marblehead," December 17, 1769, at the First Congregational Church, MDHS, item 2219.

a thriving colonial enterprise eventually rivaled England's own fishing industry, which contributed to tensions between the colonies and the imperial core. Fish merchants were not solely responsible for colonial New England's maritime commercial expansion, however. The workers and their labor are discussed in the following chapter.

3

Fishermen

> It is known to be one of the most laborious employments; that those who carry it on get to themselves but a bare subsistence.[1]

Joshua Burnham (1736–1791) was working as a skipper on the schooner *Polly* when the Revolutionary War began. His employers were Francis and John Choat, vessel-owning fish merchants in Ipswich, Massachusetts. Joshua and his younger brother Aaron were sons of Jeremiah Burnham, a middling farmer and part owner of a sawmill in the Cape Anne parish of Chebacco, near Gloucester, Massachusetts. Aaron fished commercially until he was forty years old, when his clothes and part of the vessel he was working on washed up on shore in 1782. This was a stark reminder for the Burnham family that those who went down to the sea in ships did not always return. For his part, Joshua worked as skipper and master of at least twelve vessels for various Massachusetts merchants between 1761 and 1790. Like so many other colonial maritime laborers, he died "sick from the West Indies."[2]

Eighteenth-century skippers such as Burnham were responsible for recruiting a crew, or "company," for fishing expeditions.[3] Vessel owners

[1] "Montesquieu," on "The profits accruing to the [Massachusetts cod] fishery," *Boston Evening Post*, November 28, 1763.

[2] Joshua Burnham Papers, 1758–1817, JDPL. *EVREC*, Vols. 1–2, 61, 505, 506. Daniel Vickers, *Farmers and Fishermen: Two Centuries of Work in Essex County, Massachusetts, 1630–1830* (University of North Carolina Press, 1994), 200–201, 265.

[3] See, "Petition of the Selectmen of Marblehead [To Governor Francis Bernard and the Massachusetts General Court] asking for the passage of an act relating to the Cod-fishery, with a copy of said act," 1768, MSA Collections, Vol. 66, *Maritime, 1759–1775*, 406–409.

such as Thomas Gerry and the Choats, by contrast, may have been able to indirectly influence hiring decisions, but they did not personally scour ports for crews at this time. Structural changes in the commercial fishing industry in colonial America ensured that merchants did not have to perform this duty.

This chapter investigates labor recruitment and work patterns in the colonial cod fisheries in the years leading up to the American Revolution. Again, there are two objectives. This chapter establishes who worked as cod fishermen in the colonies. It also establishes how their labor contributed to the development of the fishing industry.

As a group, eighteenth-century colonial cod fishermen were locally born, young white men who were for the most part unrelated to the skipper or the vessel-owning fish merchant they worked under. It was theoretically possible for recruiters to have hired women, older men, African Americans, Native Americans, indentured servants from abroad, and local relatives. Recruiters such as Joshua Burnham routinely chose none of these people, however. Several factors explain these recruiting decisions and this group identity.

First and foremost, a demographic shift impacted the group identity of the fishermen who worked in colonial New England's cod fisheries between the seventeenth and the eighteenth centuries. During the seventeenth century, most of the labor force along the New England coast consisted of itinerant laborers who were recruited in West Country ports to work for a specified amount of time under a master. These mobile workers were men who left New World fishing stations for their Old World homes as soon as labor contracts expired.[4] Seventeenth-century fishermen typically hailed from the Jersey Island in the English Channel. They also came from other well-known fishing depots such as Cornwall, Devon, Dorset, the coastal fishing ports of Normandy and Brittany, and ports in Wales and Ireland.[5] Such workers tended to be unmarried, young men.

[4] Vickers, *Farmers and Fishermen*, 91–92; Christine Leigh Heyrman, *Commerce and Culture: The Maritime Communities of Colonial Massachusetts, 1690–1750* (New York: W.W. Norton & Company, 1984), 212–213; and Charles E. Clark, *The Eastern Frontier: The Settlement of Northern New England, 1610–1763* (New York: Alfred A. Knopf, 1970), 24. The key difference between the migratory labor force at New England and Newfoundland was that after 1631 permanent resident fishing stations had been established in the former. This settlement contributed to the earlier expansion of local labor pools in New England.

[5] Heyrman, *Commerce and Culture*, 214; and Samuel Roads, Jr., *The History and Traditions of Marblehead*, 3rd ed. (Marblehead, MA: N. Allen Lindsay & Co., 1897), 7.

In the eighteenth century, by contrast, low mortality rates and high birth rates brought about a population boom. The total number of souls in New England expanded from 22,900 in 1650 to 170,900 in 1720, which represented nearly an eight-fold increase.[6] As prolific as this regional expansion was, fishing ports in Essex County, Massachusetts, experienced a ten-fold increase in population between 1650 and 1725.[7] Salem, Beverly, Manchester, and Gloucester grew from a combined population of 3,500 in 1690 to 16,000 in 1765.[8] Within this county-wide expansion, Marblehead, the foremost fishing port in colonial America, experienced the fastest growth. Some 1,200 people lived there in 1700.[9] By 1744, one observer estimated that 5,000 people lived and worked in Marblehead.[10] Nearby Boston's population expanded from 6,000 in 1690 to 17,000 in 1740.[11] Such growth virtually eliminated the endemic labor scarcity that plagued New England during the seventeenth century. This enabled labor recruiters to find permanent local alternatives to itinerant laborers.[12]

Gender norms also influenced labor recruiters' decisions in the colonial cod fisheries and the group identity of colonial fishermen. Historically, women played important roles in commercial fishing industries, but this was not the case in New England. Female workers, for example, processed catches on shore for the men in Scandinavian regions and in England throughout the medieval period.[13] During the eighteenth century, female settlers in Newfoundland were also mobilized for shore duty, drying, and processing catches.[14] Female labor was certainly available in colonial

[6] John J. McCusker and Russell R. Menard, *The Economy of British America, 1607–1789* (University of North Carolina Press, 1985), 103, table 5.1.

[7] Vickers, *Farmers and Fishermen*, 156.

[8] Ibid., 211.

[9] Heyrman, *Commerce and Culture*, 245.

[10] Carl Bridenbaugh, ed., *Gentleman's Progress, The Itinerarium of Dr. Alexander Hamilton, 1744* (University of North Carolina Press, 1948), 5.

[11] Gary B. Nash, *The Urban Crucible: Social Change, Political Consciousness, and the Origins of the American Revolution* (Harvard University Press, 1979), 54–55.

[12] For more on the localism involved with labor recruitment in commercial fishing in New England and England, see Vickers, *Farmers and Fishermen*, 166; and W. Gordon Handcock, *Soe longe as there comes noe women: Origins of English Settlement in Newfoundland* (St. Johns, NL: Breakwater Books, 1989), 145–216.

[13] Brian Fagan, *Fish on Friday: Feasting, Fasting and the Discovery of the New World* (New York: Basic Books, 2006), 108–110.

[14] Handcock, *Soe longe as there comes noe women*, 31–32; and C. Grant Head, *Eighteenth Century Newfoundland* (Toronto: McClelland and Stewart Limited, 1976), 218, 232. Also, see Ellen Pildes Antler, "Fisherman, Fisherwoman, Rural Proletariat: Capitalist Commodity Production in the Newfoundland Fishery" (Ph.D. dissertation, University of Connecticut, 1982), chapter 4. For the use of female labor in seventeenth-century

New England, but women were not to be found at sea or on land engaged in the fishing industry in any capacity in this region. By the time deep-sea fishing became the norm in the eighteenth century, bringing with it larger catches and the need for more workers to labor in flakeyards in ports, restrictive gender norms had hardened in New England.[15] These norms made commercial fishing a masculine endeavor and limited women to domestic work. In Newfoundland, by contrast, balanced sex ratios and offshore fishing were closer chronologically, and there was little opportunity for such norms to develop.

Race was also a factor in determining who worked in the commercial cod fishing industry. There is very little evidence of African-American participation and nothing on Native Americans. Slave labor was never an integral part of the colonial cod fishing industry, largely because New England did not maintain a large slave population.[16] To be sure, by the 1760s and early 1770s, there were a few slaves who contributed to the daily operations of this maritime enterprise. David Montgomery's slave "Cato" worked on fish merchant Thomas Davis's schooner *Swan* in the early 1770s in Beverly, Massachusetts.[17] In nearby Salem during the mid-1760s, Peter Frye forced "his Negro" to carry salt for local fish merchant Richard Derby, undoubtedly for use in curing fish taken on Derby's schooners. Frye charged Derby two shillings ten pence each day for the labor, almost exactly what white day laborers earned in the area.[18]

Newfoundland, see Peter E. Pope, *Fish into Wine: The Newfoundland Plantation in the Seventeenth Century* (University of North Carolina Press, 2004), 56–57. More work needs to be done on the role gender norms played in the colonial cod fishing industry to explain why female labor was not mobilized in New England to the extent to which it was elsewhere.

[15] Mary Beth Norton, *Founding Mothers & Fathers: Gendered Power and the Forming of American Society* (New York: Alfred A. Knopf, 1996); and Nancy F. Cott, *The Bonds of Womanhood: "Woman's Sphere" in New England, 1780–1835* (Yale University Press, 1977). In sharp contrast to eighteenth-century England, women did not work in the fields on Massachusetts's farms at that time. They were employed on colonial farms "as domestic servants, spinners, or weavers or (rarely) to pick hops, but never as farm laborers." Winifred Barr Rothenberg, *From Market-Places to a Market Economy* (University of Chicago Press, 1992), 151, footnote 6.

[16] In 1780, during the Revolution, African Americans made up only two per cent of the population in the New England colonies. By contrast, African Americans comprised forty-one per cent of the total population of the Lower South. McCusker and Menard, *The Economy of British America*, 222, Table 10.1.

[17] Thomas Davis Account Book, 1771–78, JDPL.

[18] Richard Derby Ledger, 1757–1776, JDPL.

In 1768, Captain Thomas Peach rented "his Negro Charles" to Marblehead merchant William Knight. Charles "Carried [wheel]Barrows" for Knight in May and November, most likely transporting spring and summer fare cod to and from the merchant's flakeyard.[19] Similarly, Marblehead merchant Richard Pedrick rented his "Negro" to Joshua Orne "for house" work one day in December 1776.[20] In Gloucester, Robin Boyles's "Negro" did day labor for fish merchant Daniel Rogers between 1774 and 1778. He even sold Rogers one "fish hogshead," a container for fish, in 1775 for six shillings.[21] African Americans, free or slave, could be found in all the major fishing ports. These men were surely the most exploited laborers in the commercial fishing industry. Slaves would have endured the loss of their liberty in addition to their earnings, and they would have experienced any racial antipathy that existed in an industry and a region that was mostly white.[22] Yet, such slaves were few in number.

Age also affected labor recruitment in the cod fisheries. Throughout the colonial period, commercial cod fishermen who worked Atlantic waters tended to be younger workers between the ages of ten and thirty.[23] This was primarily because the work of cod fishing proved physically demanding. Employers typically saw diminishing returns on the productive capabilities of older hands past the age of thirty.[24] There was an emphasis placed on greater and greater production capacity throughout the eighteenth century.[25] The nature of the work and the demands of productivity, then, influenced labor recruiters' decisions to concentrate on younger workers.

Kinship ties, it has been argued, further influenced recruitment in the fishing industry during the eighteenth century.[26] Boys grew into men in

[19] William Knight Account Book, 1767–1781, JDPL.

[20] Richard Pedrick Account Book, 1767–1784, MDHS.

[21] Daniel Rogers Account Book, 1770–1790, JDPL.

[22] On the hardening of racism in eighteenth-century New England, see Robert Desrochers, "Slave-For-Sale Advertisements and Slavery in Massachusetts, 1704–1781," *WMQ*, Vol. 59 (July 2002), 623–64. Also, see Ira Berlin, *Many Thousands Gone: The First Two Centuries of Slavery in North America* (Harvard University Press, 1998).

[23] Pope, *Fish into Wine*, 172; Handcock, *Soe longe as there comes noe women*, 243–248, esp. table 11.1; and Vickers, *Farmers and Fishermen*, 182–185, esp. table 7.

[24] Vickers, *Farmers and Fishermen*, 179–180, esp. figure 2.

[25] See the discussion in chapter four.

[26] According to Vickers, "the actual hiring was usually delegated to the skipper, who, as a member of the local fishing community, selected the crew from among his neighbors and

coastal communities, and it was expected they would ply their father's trade, yet Joshua Burnham took only one of his family relations, Samuel, to sea with him on the eve of the American Revolution in 1774.[27] As a labor recruiter, Burnham was not alone in this decision.

Account books belonging to colonial fish merchants are particularly useful for studying which workers were recruited for the cod fisheries. Entrepreneurs recorded accounts for their vessels, and some of these accounts include crew lists. Merchants used these lists to keep track of the amounts paid to each man. The ledgers also provide a list of the laborers who worked in the colonial fishing industry. Few such ledgers have survived, however.[28] Therefore, it is not possible to completely evaluate the role that kinship played in labor recruitment in the colonial cod fisheries with total precision.

With this caveat in mind, the surviving data indicates familial bonds actually played a limited role in labor recruitment in this maritime industry.[29] Kinship ties between vessel-owning merchants and crew

relations." Vickers, *Farmers and Fishermen*, 165. Samuel Eliot Morison writes: "A fisherman's son was predestined to the sea." Morison, *Maritime History of Massachusetts, 1783–1860*, 3rd ed. (Boston: Houghton Mifflin, 1961), 137.

[27] In the summer of 1774, skipper Joshua Burnham and his fishing company, which included George Pierce, Stephen Low, James Andrews, Samuel Burnham, Daniel Andrews, and Peter Edwards, left Ipswich, Massachusetts, in the schooner *Polly* on an expedition to the Grand Bank off Newfoundland. Joshua Burnham Papers, 1758–1817, Shipping Papers, Schooner *Polly*, 1771–1776, box 1, folder 4, JDPL.

[28] I have found ten account books belonging to late-eighteenth-century colonial fish merchants that include crew lists: William Knight Account Book, 1767–1781, JDPL; Joshua Burnham Papers, 1758–1817, JDPL; Daniel Rogers Account Book, 1770–1790, JDPL; John Stevens Account Book, 1769–75, JDPL; Richard Derby Ledger, 1757–1776, JDPL; Timothy Orne Ledgers, 1762–1767, JDPL; Miles Ward Ledgers, 1765–1772, JDPL; Thomas Davis Account Book, 1771–78, JDPL; Thomas Pedrick Account Book, 1760–1790, MDHS; and Richard Pedrick Account Book, 1767–1784, MDHS.

[29] What follows is an examination of three databases compiled by examining the ledger books of merchants involved in the commercial fishing industry. To evaluate kinship ties between vessel owners and workers, I first counted every record in which a laborer worked for a particular vessel owner. Included here are partial crew lists, skippers, and crew members. Thus, the total data set was the largest (2,114 possible berths). I then searched for instances in which worker's surnames matched the vessel owner's surnames. Servants and slaves were not calculated as kin in any of the databases.

Certain limitations complicate the data. It may have been possible for some of the workers in these data sets to have had extended family relations, including nephews or cousins, with alternate surnames. However, in the one instance in which Vickers painstakingly measured these extended kinship ties and combined them with surnames for William Knight of Marblehead, he found that fifty-five per cent of the time workers were not related to the vessel owner. Vickers, *Farmers and Fishermen*, 166. Moreover, Knight was exceptional when it came to employing his family members on fishing vessels were particularly weak.[30] Most importantly, however, there were few kinship ties between crew members and skippers, who were the primary labor recruiters responsible for physically rounding up fishing companies.[31] Such ties, then, did not guarantee work in the colonial cod fisheries.

There are several explanations for these sparse kinship ties. First and foremost, skippers and merchants did not want to subject their family members to the rigors and dangers of deep-sea fishing. Commercial fishing became more dangerous in the eighteenth century with the shift from shallops and inshore fishing waters to schooners and deep-sea waters. Deep-sea fishing took workers far from the safety of ports and natural harbors, which made the stormy weather prevalent on the North Atlantic that much more of a factor.

Deep-sea cod fishing also had a reputation for being exceptionally physically demanding. Christopher Prince was born in Kingston, Massachusetts, in 1751. His father, who was a seaman-turned-farmer, knew that commercial fishing was by no means an easy way to make a living. Indeed, that is precisely why he allowed his son to join a cod fishing company in the first place. According to Prince, who ended up not pursuing a career in commercial fishing, his father "had hopes from the commencement by [Prince] going a fishing would wean [him] from pursuing" a life at sea.[32] Fathers like Prince's wanted a less rigorous and less risky career for their sons.

members. If ever there were a case to be made for kinship ties playing a decisive role in labor recruitment a majority of the time, Knight would have been the best illustration.

In addition, assembling crew lists from merchant ledgers and shipping papers can be problematic. At times, incomplete or illegible ledgers made it impossible to compile lists for certain fares. Such fragmentary crew lists were only used in those cases when this was warranted. Partial crew lists were included along with complete crew lists in the first two databases where I test whether skippers or vessel owners had more of a say in labor recruitment to get at as much of this picture as possible. Only full crew lists were used in evaluating kinship ties between crew members.

[30] Only sixty-four of the total 2,114 possible berths, just three per cent, were occupied by workers with kinship ties to vessel owners. This database is larger than the one in footnote 32 because partial crew lists were used here and skippers were calculated.

[31] Fishermen who shared the skippers' surnames occupied only 193 out of 1,673 possible berths, or twelve percent. Put another way, eighty-eight per cent of the time labor recruiters chose non-family members to go to sea with them based on the above calculation of surnames. If we were to double the twelve per cent to roughly estimate female kinship ties, then the former total would be seventy-six per cent.

[32] Michael J. Crawford, ed., *The Autobiography of a Yankee Mariner: Christopher Prince and the American Revolution* (Washington, DC: Brassey's, Inc., 2002), 18.

Additionally, the population increase discussed in the beginning of the chapter made it possible for labor recruiters to avoid hiring family labor or migrant workers. More workers, ready at hand meant skippers could choose who went to sea with them, and who did not. Increased local labor pools also meant that vessel-owning fish merchants could afford to pick and choose when it came to their input into who worked on their vessels. Merchants and skippers did not want their family members risking life and limb on the Atlantic Ocean. For these reasons, then, the labor recruiters in this study were very particular about who could work in the fishing industry, and by-and-large they chose young, white, male, non-family members to go to sea with them.

This is not to say that kinship ties were completely absent on commercial fishing vessels. Maritime laborers beneath the skipper often worked at sea with family members in the same vessel. Poorer mothers and wives may have wanted their sons and husbands to work on separate vessels to lessen the risk of losing multiple loved ones in a storm or work-related accident. However, the same demographic shift that enabled labor recruiters to be selective in their hiring practices also worked against laborers by generating underemployment. There were simply more men than jobs throughout New England communities in the eighteenth century.[33] As a result, workers grabbed their out-of-work relation whenever a skipper required hands. For example, Benjamin Peters and his son, Benjamin Jr., joined the company of the Salem schooner *Lucretia* for the second and third fares in 1765.[34] Jeremiah and Jonathan Foster likewise worked on the Beverly schooner *Swallow* for two fares in 1771.[35] James and William Jones worked together on the Gloucester schooner *Britannia* for three fares in 1772.[36] John and James Bowden joined the company of the Marblehead schooner *Barnett* on a fall fare in 1773.[37] And James Andrews and Daniel Andrews shipped together on the *Polly* for a summer fare out of Ipswich in 1774.[38] Out of 1,608 individual berths recorded between 1752 and 1775, slightly more than

[33] Vickers, *Farmers and Fishermen*, 187–188; and Robert A. Gross, *The Minutemen and Their World* (New York: Hill and Wang, 1976), 78–79, 86–89. Gross describes the farming community of Concord, Massachusetts, as "an economy of increasing scarcity in an environment of spreading blight" on the eve of the American Revolution. Ibid., 88.

[34] Miles Ward Ledgers, 1765–1772, JDPL.

[35] Thomas Davis Account Book, 1771–78, JDPL.

[36] John Stevens Account Book, 1769–75, JDPL.

[37] William Knight Account Book, 1767–1781, JDPL.

[38] Joshua Burnham Papers, 1758–1817, Schooner *Polly*, 1771–1776, box 1, folder 4, JDPL.

half (fifty-four per cent) of the time at least two crew members shared the same surname. Thirteen per cent of the time there were two kinship ties between the crew members. Three per cent of the time there were three such ties on board. As a result, horizontal social networks among fishing crews were denser than the vertical networks between crew members and those responsible for labor recruitment in the industry.

The work that these men performed changed over the course of the colonial period. Offshore fishing operations only started to become the hallmark of the cod fisheries in the eighteenth century. For much of the seventeenth century, the fishing industry in colonial New England was prosecuted on an inshore basis.[39] Colonists typically worked along the coast between Cape Cod and the province of Maine. Smaller, three-man crews, consisting of a foreshipman, midshipman, and steersman, worked close to shore and near homeports in shallops. Shallops were open-decked, doubleended, single or double-masted vessels averaging twenty-five feet in length and six tons burden.[40] Catches were taken to shore, where the men dressed and salted the fish on stages, enclosed wooden wharves. Workers then spread the filets on fish flakes, or open-air wooden platforms, for drying. At times, a specialist known as a shoreman was taken along to assist the fishermen with processing the catch on land.

Schooners such as the *Polly* began replacing smaller shallops around the turn of the eighteenth century.[41] These hardy craft revolutionized

[39] Vickers, *Farmers and Fishermen*, chapter 3, esp. 122–126. Pope provides a concise definition of these two fisheries: "Inshore fisheries are prosecuted from shore in day-to-day voyages by open boats on fishing grounds close to the coast; offshore fisheries are prosecuted from ships voyaging for weeks at a time to fishing banks, which may be days or even weeks sailing from land." Pope, *Fish into Wine*, 14.

[40] For more on shallops, see Fagan, *Fish on Friday*, 246n; and Vickers, *Farmers and Fishermen*, 121, esp. footnote 83. The most systematic examination of schooners is Basil Greenhill, *The Merchant Schooners*, 2nd ed. (Annapolis, MD: Naval Institute Press, 1988). Also, see Vickers, *Farmers and Fishermen*, 145–146.

[41] The origins of the schooner are as yet obscure. John J. Babson, a Gloucester, Massachusetts, native, first put the place and date of the schooner's birth at Gloucester, 1713, owing to the unsubstantiated claim of Captain Andrew Robinson, another native of Gloucester. John J. Babson, *History of the Town of Gloucester, Cape Ann, Including the Town of Rockport* (Gloucester, MA: Peter Smith Publisher, Inc., 1972; orig. pub. 1860), 248–255. In succession, McFarland, Robert Greenhalgh Albion and Jennie Barnes Pope, Innis, and Vickers followed Babson's lead. Raymond McFarland, *A History of the New England Fisheries* (New York: D. Appleton & Company, 1911), 82; Robert Greenhalgh Albion and Jennie Barnes Pope, *Sea Lanes in Wartime: The American Experience, 1775–1942* (New York: W.W. Norton and Company, Inc., 1942), 20; Harold Adams Innis, *The Codfisheries: The History of an International Economy* (Yale University Press, 1940), 166; and Daniel Vickers, *Young Men and the Sea: Yankee Seafarers*

and forever changed the commercial fishing industry in the colonies. They were typically twenty-to-100 tons in burden, "two-masted ocean-going craft – thirty-five to sixty-five feet in length and fully decked with a raised forecastle or cabin abaft."[42] Unlike shallops, schooners could handle extended periods in offshore waters, and they could even make transatlantic passages. Access to waters far from shore was crucial to greater success in commercial fishing. For, as Brook Watson, "a Merchant" acquainted with "the Fisheries of North America," testified before Parliament in 1775, "larger Fish" were taken "in deeper water."[43]

Whereas seventeenth-century fishing vessels were typically manned by three, mostly experienced fishermen, more workers could fit on schooners, which averaged seven-man crews.[44] Larger crew sizes brought increased variation in age and skill, and by the middle of the eighteenth century there was a clear hierarchy in the commercial fisheries. At the top of the labor hierarchy stood the skipper. He was responsible for recruiting crews, choosing where to fish, navigating the vessel, fishing, and

in the Age of Sail (Yale University Press, 2005), 75. William Avery Baker, however, forcefully contends: "The story of the supposed invention of the schooner in Gloucester about 1713 need not be repeated here, for the schooner rig . . . existed before that date." William Avery Baker, "Vessel Types of Colonial Massachusetts," in *Seafaring in Colonial Massachusetts: A Conference Held by the Colonial Society of Massachusetts, November 21 and 22, 1975* (Boston: The Society; distributed by University Press of Virginia, 1980), 19. For his part, Peter Pope has referred to the ketch, "a small, seaworthy, beamy, flush-decked, double-ended vessel" developed in the mid-seventeenth century, as "a protoschooner." Pope, *Fish into Wine*, 153. Greenhill, the foremost expert, argues that "as far as it is possible to tell," the brigantine evolved into the schooner over "the course of the eighteenth century." Greenhill, *The Merchant Schooners*, 6. However, he notes that schooners may have derived from similar Dutch models dating from the turn of the seventeenth century. For similar arguments that the schooner rig was not first designed in America, see William M. Fowler Jr., *Rebels Under Sail: The American Navy During the Revolution* (New York: Charles Scribner's Sons, 1976), 265n; and Howard I. Chapelle, *The Search for Speed Under Sail, 1700–1855* (New York: Norton, 1967), 11. A comparative Atlantic history of shipbuilding would probably resolve the debate.

42 Vickers, *Farmers and Fishermen*, 145. By comparison, shallops averaged six tons burden in 1676. Ibid., 121n.

43 *AA*, Series, 4, Vol. 1, 1642.

44 Boatmasters, or steersmen, and midshipmen tended to be more experienced, skilled laborers, while foreshipmen were usually unskilled boys and green hands. These early workers were supported on shore by skilled laborers such as the splitter, header, and salter. Vickers, *Farmers and Fishermen*, 123–125; and Pope, *Fish into Wine*, 176–177, esp. table 9. According to one seventeenth-century observer at Newfoundland, "the boats' masters, generally, are able men, the midshipman next, and the foreshipmen are generally striplings." Quote taken from Head, *Eighteenth Century Newfoundland*, 3. The average crew size was 7.4 men. This figure was calculated from the account books cited throughout this chapter.

keeping a count of the fish caught.[45] The sharesmen were the older, more experienced hands on board the schooner. Their primary duties revolved around catching cod and helping to process the catch. The cuttails were the younger, inexperienced greenhands. They were mainly tasked with cutting bait, baiting hooks, and processing the catch, although they did help out with fishing. Each worker assisted in sailing.

Schooners also brought about a shift to a two-part production process, the combination cure, which the Dutch first developed in the late fourteenth century.[46] In the first stage of this process, fishermen sailed to deep-sea banks where handlines were lowered and cod was hauled in. While still at sea, fishermen would behead, gut, bone, and lightly salt the cod on board the schooners' ample decks. During the second stage of the process, the crew returned to their homeport, and the cure was completed by air-drying the catch on fish flakes in flakeyards. Schooners were required for this production process. Shallops did not enable fishermen to reach offshore banks, nor did they allow fishermen to work at sea for extended periods of time. In addition, shallops did not enable men to partly process catches on deck while still on the ocean.

Commercial fishing in the colonies during the eighteenth century was further separated from the earlier period by the rarity of winter expeditions. It was not uncommon for colonial cod fishermen in the seventeenth century to work close to shore throughout the year.[47] Unlike seasonal migratory species of fish such as herring, cod perennially inhabit

[45] The skipper's duties on land and at sea are discussed at length in Christopher Paul Magra, "The New England Cod Fishing Industry and Maritime Dimensions of the American Revolution" (Ph.D. Dissertation, University of Pittsburgh, 2006), chapter 1.

[46] Pope, *Fish into Wine*, 27; and Vickers, *Farmers and Fishermen*, 149, 150. Vickers refers to the combination cure as "a new cure." Ibid., 149. However, Pope notes that "similar processes had long been in use on the other side of the Atlantic." Pope, *Fish into Wine*, 27. For the employment of the combination cure among eighteenth century English settlers in Newfoundland, see Head, *Eighteenth Century Newfoundland*, 72–73. Head suggests that British fishermen learned the cure from French fishermen in the Newfoundland region. The combination cure was not dominant in Newfoundland in the eighteenth century, which partly accounts for the predominance of Newfoundland cod in European markets. For more on the Dutch cure, see R.W. Unger, "The Netherlands Herring Fishery in the late Middle Ages: The False Legend of Willem Beukels of Biervliet," *Viator*, Vol. 9 (1978), 335–356. Although the cure was developed for use in Dutch North Sea herring fisheries, it was extended to other types of fish and used in the taking of cod in Atlantic waters around medieval Ireland. See Maryanne Kowaleski, "The Western Fisheries," in "Fishing and Fisheries in the Middle Ages," in David J. Starkey, Chris Reid, and Neil Ashcroft, eds., *England's Sea Fisheries: The Commercial Sea Fisheries of England and Wales Since 1300* (London: Chatham Publishing, 2000), 27.

[47] Vickers, *Farmers and Fishermen*, 123.

the same area. The prevalence of winter fishing changed, however, due to ever-present climatic risks and the transition to offshore fishing.[48] Macro-level alterations in the Atlantic economy also diverted workers into winter trade voyages south to slave plantations during the winter months so that, during the eighteenth century, only a minority of the fishermen conducted expeditions during the winter months, and those who did were rarely successful.[49]

Benjamin Bangs, for example, worked on cod fishing expeditions out of Eastham, Massachusetts. He attempted to work the inshore waters off Cape Cod in February 1743. On this expedition, he encountered "a hard NE snow storm" and a "Cold NW wind." Bangs observed that it was a "tuff time" and that he and the other fishermen could "Catch but 3 Codfish." On the return trip, he encountered a "NE storm of Rain and snow" and referred to the entire expedition as "a broken voyage."[50]

Similarly, the foolhardy few who attempted to reach deep fishing waters during the winter learned their lesson the hard way. In 1760, Robert Hooper Jr., a fish merchant from Marblehead, petitioned the Massachusetts General Court to be reimbursed for a rescue mission. A "Number of People" were working the waters around Sable Island off the coast of Nova Scotia when they were "cast on Shore on said Island and in Danger of perishing by the Severity of the Winter, unless relieved."[51] A dramatic instance such as this served as a stern warning to fishing communities up and down the New England coast.

The majority of eighteenth-century Massachusetts fishermen went fishing to offshore banks in the spring, summer, and fall. In the cold winter months, fishermen typically worked for their same merchant employers, on the same vessels, on trade voyages to slave plantations in the Southern mainland and West Indian colonies.[52] Others did day labor in their employer's flakeyard, processing the fall catch.[53]

[48] The period 1300–1850 witnessed unusually cold temperatures and harsh winters. Brian M. Fagan, *The Little Ice Age: How Climate Made History, 1300–1850* (New York: Basic Books, 2001).

[49] The shifting direction of the Atlantic cod trade is discussed in chapter four.

[50] Benjamin Bangs Diary, 1742–1765, MHS.

[51] *JHRM*, Vol. 38, Part II, 278.

[52] See the discussion in chapter four.

[53] Magra, "The New England Cod Fishing Industry and Maritime Dimensions of the American Revolution," chapter 1.

Fishing schooners were generally taken out of homeports in March and April to catch spring fare on the first expedition of the year.[54] In 1753, for instance, Timothy Orne gave orders for John Felt, a skipper who was working as captain on a trade voyage, to "Be sure to get Ready to Come away by the Last of January if Possible that so you may be at home timely for The Spring Sable [Bank] Fare, which is of more Consequence than the Maryland Voyage."[55] Leaving the Southern colony between January and February would have allowed Felt and his crew enough time to make it home, get outfitted for fishing, sail to the banks, and catch the first fish of the season off Sable Island.

Between the spring and fall months, eighteenth-century colonial fishermen frequented the offshore waters between Maine and Nova Scotia.[56] The Sable Island Bank, located southeast of Nova Scotia, 300 miles east of Maine's coast, or four days sailing time for Massachusetts's fishermen, was the most popular fishing waters in this region.[57] Raymond McFarland describes the Sable Island Bank as "elliptical in form with a length of 156 miles and a width of 56."[58] Fishermen could travel here whether it was the spring fare, the summer fare, or the fall fare that ended in October or November.

During the summer months, colonial fishermen like the crew of the *Polly* typically headed to the "eastward" and the largest known feeding and spawning waters of Atlantic cod – the Grand Bank.[59] Eighteenth-century fishermen did not work the bank during the fall, winter, or spring

[54] Deborah C. Trefts asserts, "[T]he cod season began in mid-winter (February) and continued virtually year-round." Deborah C. Trefts, "Canadian and American Policy Making in Response to the First Multi-species Fisheries Crisis in the Greater Gulf of Maine Region," in Stephen J. Hornsby and John G. Reid, eds., *New England and the Maritime Provinces: Connections and Comparisons* (McGill-Queen's University Press, 2005), 209. I have seen no evidence of fishermen leaving in February to go to the banks.

[55] Schooner *Molly*, box 7, folder 10, 1751–57, Timothy Orne Shipping Papers, JDPL.

[56] Indeed, one of the fishing grounds off the coast of Maine was named "Marblehead Bank." McFarland, *A History of the New England Fisheries*, 12.

[57] Trefts, "Canadian and American Policy Making in Response to the First Multi-species Fisheries Crisis in the Greater Gulf of Maine Region," 206–211; Vickers, *Farmers and Fishermen*, 148–149; and McFarland, *A History of the New England Fisheries*, 14.

[58] McFarland, *A History of the New England Fisheries*, 14.

[59] Joshua Burnham Papers, 1758–1817, Shipping Papers, Schooner *Polly*, 1771–1776, box 1, folder 4, JDPL. For other examples of eighteenth-century Massachusetts's schooners traveling to the Grand Bank in the summer months, see George Stevens Logbooks, 1768–1774, MHS; schooner *Ruby*, 1788; schooner *St. Peter*, 1793; microfilm 91, reel 68, JDPL; and the log of the schooner *Nancy* 1795–96, microfilm 91, reel 3, JDPL.

months, because of dangerous temperatures, winds, and seas, combined with the prevalence of fog and icebergs in the region.[60] McFarland notes that the bank "is roughly triangular in shape, one side facing N.N.E., another S.W., and the third about E. by S. North and south it extends from below the parallel of 43° to beyond that of 47°; its width is between the meridians of 48° and 54°, giving it an area of 37,000 square miles, or more than that of the state of Indiana."[61] At this point in time, no other part of the Atlantic Ocean teemed with cod like the Grand Bank.

The Grand Bank was located southeast of Newfoundland 600 miles east of Maine's coast. After a passage of ten days' sailing time, during which the younger cuttails probably cut mackerel and clams into bait while the experienced sharesmen swapped stories, fishermen who reached this majestic stretch of the ocean dropped anchor.[62] The men would have taken soundings before breaking the water with handlines, lead weights, bait, and hooks. John Thistle, skipper of the Beverly schooner *St. Peter*, had the crew take a sounding on September 16, 1793. They reported they were in seventy-five fathoms "on the grand Bank." Later that day, the *St. Peter*'s crew reported a depth of "60 fathoms of warter."[63] Skipper John Groves of the Danvers schooner *Nancy* and his crew came to anchor on the Grand Bank at latitude 45°N 7′ and longitude 52°W 7′

Also, see the statement of Alexander Coffin, a Massachusetts fish merchant, in L.H. Butterfield, ed., *The Adams Papers: Diary & Autobiography of John Adams, Volume III, Diary 1782–1804 and Autobiography through 1776* (New York: Atheneum, 1964), 74.

[60] According to Pope, the bulk of fishing activities were conducted off Newfoundland between May and July. Pope, *Fish into Wine*, 21–29. Also, see Vickers, *Farmers and Fishermen*, 104, 116. One observer recorded "Lands of Ice" floating around Newfoundland between March-April 1726, and further noted that this made it "very dangerous for those that are unacquainted." Quoted in Head, *Eighteenth Century Newfoundland*, 73. Head notes that a majority of the region's gale-force winds come between December and March. At the same time, the region experiences its highest number of days with precipitation and its lowest temperatures. By contrast, June, July, and August typically maintain low precipitation, slight winds, and warm temperatures. Ibid., 43.

[61] McFarland, *A History of the New England Fisheries*, 14–15.

[62] The average sailing time from Beverly, Massachusetts, to the Grand Bank in the late eighteenth century was taken from the log of the schooner *St. Peter*, 1793, Ship's Log Books, microfilm 91, reel 68, JDPL; the log of the schooner *Nancy*, 1795–1796, Ship's Log Books, microfilm 91, reel 3, JDPL; and the log of the schooner *Lark*, 1771, George Stevens Logbooks, 1768–1774, MHS. By comparison, it took migratory fishing fleets an average of five weeks to travel from their ports in England's West Country to Newfoundland in the eighteenth century. Ian K. Steele, *The English Atlantic 1675–1740: An Exploration of Communication and Community* (Oxford University Press, 1986), 82.

[63] Log of the schooner *St. Peter*, 1793, Ship's Log Books, microfilm 91, reel 68, JDPL.

in July 1795. They sounded a depth of thirty-six fathoms.[64] By recording ocean depths, fare after fare, year after year, skippers accumulated knowledge about where fish could be caught. Such maritime knowledge could separate the successful fares from the "broken voyages."

Burnham and company worked the Grand Bank from Sunday, June 26, 1774, to Friday, July 15, 1774. Joshua Burnham took 927 cod; Samuel Burnham, 2,000; Peter Edwards, 1,913; George Pierce, 1,797; Stephen Low, 1,668; James Andrews, 1,573; Daniel Andrews, 1,384. In all, the seven-man crew took in a catch of 11,262 cod.[65] This figure corresponds to an average catch-rate-per-man of 1,609. Assuming that 54.5 fish made up a single quintal, then this catch rate in quintals was thirty quintals per man.[66] This figure was low for the period. Between 1675 and 1775, there was a seventy-five per cent increase in the average catch-rate-per-man in the New England cod fishing industry, from forty-five to seventy-nine quintals, as a result of the transition to schooners and larger crews, which enabled fishermen to access larger fish in offshore waters.[67] The fish count for Burnham and his crew may have also been low in comparison to typical catch-rates-per-boat for the Newfoundland deep-sea fishery. Using the estimate of 54.5 fish per quintal, the catch rate for the *Polly* was 207 quintals.[68]

Deep-sea fishing had a reputation as being very physically demanding. A Boston writer referred to it as "one of the most laborious employments."[69] Christopher Prince described the work as "the most trying of any employment."[70] Fishermen worked a grueling, near continuous schedule for as long as they stayed on the banks.[71] Daniel Vickers

[64] Log of the schooner *Nancy*, 1795–1796, Ship's Log Books, microfilm 91, reel 3, JDPL.

[65] Schooner *Polly*, 1771–1776, box 1, folder 4, Joshua Burnham Papers, box 1, Shipping Papers, 1758–1817, JDPL. By comparison, the earlier English migratory fishery at Newfoundland averaged over 20,000 fish on expeditions lasting 56–70 days. Pope, *Fish into Wine*, 23.

[66] For the calculation of 54.5 fish per quintal, see the discussion in chapter four.

[67] Vickers, *Farmers and Fishermen*, 100, 154, esp. table 4.

[68] According to Pope, "a catch of 200 quintals was considered normal" in the seventeenth century in-shore fishery at Newfoundland. Pope, *Fish into Wine*, 37, table 1. Presumably, the offshore fishery of the eighteenth century witnessed an increase in "normal" catch rates. Indeed, William B. Weeden estimated that during this later period, "a good catch was 500 or 600 quintals." William B. Weeden, *Economic and Social History of New England, 1620–1789*, reprint ed., Vol. 2 (New York: Hillary House Publishers, Ltd., 1963), 751.

[69] *Boston Evening Post*, November 28, 1763.

[70] Crawford, ed., *The Autobiography of a Yankee Mariner*, 16.

[71] John Groves and the crew of the schooner Nancy worked on the Grand Bank for forty-six days between July and August 1795. Log of the schooner *Nancy*, 1795–1796, Ship's Log

estimates men worked eighteen to twenty hours per day in the inshore fisheries, and "most of the day and frequently in turns throughout the night" in the offshore fisheries.[72] After "ten weeks" on the Grand Bank in the summer trying to catch fish, Christopher Prince "had not one clean garment to wear."[73]

At bottom, labor in the fishing industry involved man's struggle and symbiotic relationship with nature. On the one hand, fishermen fought to balance themselves on board wooden vessels caught in rolling seas. They tacked one way and another to make headway in the face of contrary winds. Fishermen also attempted to stave off cold, damp weather with thick clothing, yet, at the same time, fishermen moved with the rhythms of nature. As seasons changed, work patterns were altered. As water temperatures altered, and cod stocks shifted, fishermen moved to different zones. In these ways, fishermen would have fought with and learned from nature on their sea voyages.

Fishermen encountered all sorts of foul weather while working on the offshore banks of the Northwest Atlantic littoral that made their work that much harder. Joshua Burnham, now working out of Gloucester for Daniel Rogers on the schooner *Ruby*, was fishing "on the Grand Bank" on Friday, September 19, 1788. Two days later, the Gloucester crew "hove up our anchor" and headed for Cape Anne. At 2 a.m. Tuesday, September 23, ninety-one leagues west of the Grand Bank, the *Ruby* was hit by "a hard Gale of wind at S.S.E" that forced the crew to reef sails. By 10 a.m., there was "wind at S. By W. hard Gale." Twenty-four hours later, "the wind abated," and they headed "S.B.W. under 4 Reefed fore Sail." According to the next observation in Burnham's log, they were only eighty-two leagues west of the Grand Bank following the gale-force winds.[74] John Thistle observed "thick weather" when he and his Beverly crew were "on the grand Bank" in mid-September 1793. On the return to Massachusetts that same fare, Thistle "spoke with a Cape Ann man from the grand Bank. He had lost all of his cables and anchors" in

Books, microfilm 91, reel 3, JDPL. Skippers George Stevens and Thomas Woodberry, together with their respective crews, worked on the Grand Bank for fifty-two days between June and August 1770. Log of an unnamed schooner, 1774, George Stevens Logbooks, 1768–1774, MHS.

[72] Vickers, *Farmers and Fishermen*, 122, 150.

[73] Crawford, ed., *The Autobiography of a Yankee Mariner* 15. After fishing the Grand Bank and along the New England coast for two years in 1764 and 1765, Prince decided "to follow the sea for a living the remainder of my days... but not a fishing." Ibid., 18. He went to work instead as a full-time merchant mariner.

[74] Log of the schooner *Ruby*, 1788, Ship's Log Books, microfilm 91, reel 68, JDPL.

a sudden storm, presumably related to the "thick weather."[75] In 1769, George Stevens left Beverly on a spring fare to the Sable Island Bank. Near latitude 42°N 51′ and longitude 59°W 40′, southeast of Sable Island, on Wednesday, March 4, Stevens observed that "this 24 [hours] past fine winds and Cloudy weather with Some Squalls of Snow." On Thursday, March 5: "This 24 past moderate breeze of wind and Clear weather." Then, near latitude 43°N 21′ and longitude 59°W 10′, he encountered "rainy weather." Later, on the return trip back to Beverly, at latitude 39°N 23′ and longitude 68°W 47′, he observed, "[T]hese 24 [hours] past hard winds so that we was obliged to Scud."[76] Afterward, Stevens and his company set sail from Beverly to Banquereau Bank, located off the east coast of Nova Scotia, on a first fare in April 1774. The men only caught "some" cod there, however, and they headed for the Grand Bank at the end of the month. En route, Stevens observed "fresh gales with plenty of rain and Cold Weather." At another time, on a fall fare to Banquereau Bank in mid-August, he saw "foggy, misty weather."[77] One horrified English migratory fisherman attempting an early trip to the Grand Bank in March 1670 observed arctic winds and waters that turned his vessel into "a lump of ice.... the water freezes as soon as it comes on decks, not for our lives able to loose a knot of sail, all things are so frozen.... God help us, we are very fearful."[78] As late as May 1766, Newfoundland merchant Samuel Prince encountered "nothing but hard gales & contrary winds & snow storms the chief of our passage" from Boston to St. John's. "For my part," he wrote to Boston merchant Isaac Clarke, "I began to think we had mistaken the season of the year & left Boston in January instead of May, so much for our passage."[79]

Such bad weather contributed in different ways to the work-related dangers fishermen faced. Several newspapers in New England reported in 1766 the effects of storms at sea on various sailors and fishermen. Included in these accounts was the description of "the Bodies of two Men" that had washed up on the shore off Sable Island near the wreck of a schooner, "which by their Dress appeared to be Fishermen."[80] The Burnham family could surely sympathize. A Worcester, Massachusetts, newspaper writer reported hearing from Boston of a "violent storm of

[75] Log of the schooner *St. Peter*, 1793, Ship's Log Books, microfilm 91, reel 68, JDPL.
[76] Log of an unnamed schooner, 1769, George Stevens Logbooks, 1768–1774, MHS.
[77] Log of an unnamed schooner, 1774, George Stevens Logbooks, 1768–1774, MHS.
[78] Quote taken from Head, *Eighteenth Century Newfoundland*, 2.
[79] Samuel Prince to Isaac Clarke, Trinity, Newfoundland, June 21, 1766, MHS.
[80] *Boston Evening Post*, June 23, 1766; *Boston News-Letter and New-England Chronicle*, June 26, 1766; *Connecticut Courant*, June 30, 1766.

the 19th" of November 1783, "Since which several other dead bodies have been taken up, which appear to be fishermen."[81] Shipwreck could also occur as a result of collision, usually when larger vessels failed to spot fishing vessels moored on the banks at night.[82] Accounts of drowned fishermen were never big news, meriting only a few sentences in the back pages of newspapers, yet such accounts were not uncommon.[83] Men were also reported "lost" or "drowned" while working on the offshore banks in vital records pertaining to deaths.[84] Between January 1768 and June 1770, "Twenty-four" fishing vessels from Marblehead alone, along with "One hundred and seventy Men and Boys," were "lost at Sea."[85] Fish merchants may have taken certain risks investing in commercial fishing expeditions. However, workers risked life and limb – especially in bad conditions.

Entire crews handlined for cod while working offshore. According to Christopher Prince, "when we come across a school of fish, every one is anxious to get as many as any of his shipmates." Each man dropped two lines over the vessel's bulwark and waited for the lead weights to carry the hooks and bait close to the ocean bottom. They then anxiously lingered for a bite, and their patience was not always rewarded. The frustration of fishing for cod unsuccessfully can be felt in Prince's words on the subject. He worked the Banquereau, or "Quero" Bank, for two years, and later wrote:

[81] *Massachusetts Spy; Or, Worcester Gazette*, December 4, 1783.

[82] See, for example, *Boston Gazette, and Country Journal*, July 18, 1763; *Boston Evening Post*, July 18, 1763; and *Newport Mercury*, July 25, 1763.

[83] In addition to the accounts in this paragraph, see *Boston Post Boy*, September 10, 1764; and *New-London Gazette*, October 8, 1773.

[84] For Aaron Burnham's drowning in 1782, see footnote 2 above. Also, see the death record for John Chin, Marblehead fisherman. Chin was "washed over[board] in [a] spring fare" in 1745. *EVREC*, Vol. 2, 515. Similar examples of Marblehead fishermen's deaths for a later period include Charles Chadwick, who "Sailed in August last in a Schooner belonging to Mr. Samuel Knight for the Grand Bank and never been heard of Since" in 1815. John Courtis was also on Knight's schooner. Ibid., 524. One of Chadwick's relations, another Charles, "Drowned on Grand Bank, Schooner *Senator*," in 1846. Ibid., 512. Isaac Collyer was "lost at Sea Coming from Grand Bank last September (Schooner *Susan*), in 1825. Ibid., 521. William Cole "and company, lost going to Grand Bank (Schooner *Panther*)," in 1830. Ibid., 520. William Chambers was "Lost Overboard out of the Schooner (*Friendship*), Samuel Thompson, Skipper, on Grand Bank" in 1831. Ibid., 513. John William Caswell was "one of the Crew of the Schooner *Ocean* lost on the passage from the Grand Bank" in 1840. Ibid., 511. One of his relations, Thomas P. Caswell, "drowned on Grand Bank, Schooner *Zela*," in 1846. Ibid., 512.

[85] *By the Honorable Thomas Hutchinson*... (Boston: Richard Draper, 1770), Early American Imprints, Series I: Evans #42125.

> At one time we came among a large quantity of fish, and they were hauling of them in, almost every one but myself, without any intermission, and I could not get one. If I felt a bite, it was only to rob my hooks of their bait, and sometimes I would hook one and get it near the top of the water and then it would break off. After experiencing many of these trials, which I bore for some time with Christian patience, I at last gave way and for the first time in all my life I uttered a profane word.[86]

A more fortunate fisherman hauled up his catch using wooden hand-frames and the bulwark for leverage. A single cod could weigh from five to a hundred pounds and, in rare instances, could exceed two hundred pounds. Cod tongues were taken out and separated for each of the more experienced men. Younger crew members cut the tails of fish with different marks, such as stumps, swallows, doubles, or singles, to indicate their catches. The marked and de-tongued cod was then tossed into a container nearby.[87]

When the container began overflowing, men stopped hauling cod out of the ocean and began processing the catch. Part of the crew stood on the schooner's deck splitting cod bellies open, ripping out guts, cutting off heads and tails, and then removing spines. The marked tails were presumably returned to their owner for the purposes of counting. Fishermen carefully removed the cod's liver to be converted into a marketable oil. Other parts of the entrails were made into bait. Workers, most probably the younger cuttails, also had the unsavory job of going down into the schooner's dank bowels to receive the filets. They lightly salted the filets and stacked them in the hold. This light salting was, again, part one of the two-stage curing process. The entire crew then repeated the process for as long as they were physically able or until the schooner's hold had been filled.[88]

[86] Crawford, ed., *The Autobiography of a Yankee Mariner*, 16, 18.

[87] Vickers, *Farmers and Fishermen*, 150–151, 173–174. For the lower-limit for cod weight, see ibid., 123–126. For the middling weight, see Head, *Eighteenth Century Newfoundland*, 5. For the upper limit of cod weight, see Innis, *The Codfisheries*, 4. Pope notes that the technology used in handlining for cod was medieval with the exception of the wooden hand-frames for winding the line. Pope, *Fish into Wine*, 25n.

[88] Pope, *Fish into Wine*, 22–29, esp. plate 2; Head, *Eighteenth Century Newfoundland*, 72–74; Innis, *The Codfisheries*, 48. Dories were not widely used in the cod fishing industry until the nineteenth century. See Andrew A. Rosenberg, W. Jeffrey Bolster, Karen E. Alexander, William B. Leavenworth, Andrew B. Cooper, and Matthew G. McKenzie, "The history of ocean resources: modeling cod biomass using historical records," *Frontiers in Ecology and the Environment*, Vol. 3, No. 2 (2005), 85, figure 1. The cod trap displaced handlining in the late nineteenth century.

On Saturday, July 16, the crew of the *Polly* weighed anchor and "sailed from the Grand Bank for home."[89] Assuming they encountered good weather, their return trip would have been one week longer than their voyage out, seventeen days on average, due to contrary westerly winds.[90] In total, then, fishermen traveling to and from the Grand Bank in the late eighteenth century could expect almost a month of sailing time.

Time was a crucial element in processing raw fish and getting it to market. The pressures of the marketplace governed early modern fishermen every bit as much as urban industrial workers.[91] Consumers wanted properly cured fish, and the first fish of the season commanded the highest price. The quicker the combination cure could be applied to fish, the better the results. This put pressure on fishermen to catch, process, pack, and transport cod to homeports as quickly as possible. For example, on November 14, 1795, the schooner *Nancy* weighed anchor and "sailed from Grand Bank Bound for Beverly [Massachusetts]." Skipper John Groves prayed, "God send us a *fast passage* to our Destined port." Despite the weather and work-related accidents that could hinder his passage, he did not pray for *safe* passage home. Even late in the year, speed mattered most to fishermen.[92] The catch then had to be fully cured before

[89] Schooner *Polly*, 1771–1776, box 1, folder 4, Joshua Burnham Papers, 1758–1817, box 1, Shipping Papers, JDPL.

[90] The average sailing time from the Grand Bank to Beverly, Massachusetts, in the late eighteenth century was taken from the log of the schooner *St. Peter*, 1793, Ship's Log Books, microfilm 91, reel 68, JDPL; the log of the schooner *Nancy*, 1795–1796, Ship's Log Books, microfilm 91, reel 3, JDPL; the log of the schooner *Lark*, 1771, George Stevens Logbooks, 1768–1774, MHS; and the log of an unnamed schooner, 1774, George Stevens Logbooks, 1768–1774, MHS. In comparison, the West Country migratory fishing fleets routinely faced homeward bound travel times of three weeks. Steele, *The English Atlantic*, 82. In sum, it took the fleets that sailed out of England longer to reach Newfoundland than it did to sail home. The opposite was the case for New England fishermen, due primarily to wind speeds and ocean current direction.

[91] Vickers, *Farmers and Fishermen*, 117, 119–123, 127; Pope, *Fish into Wine*, 171–172; and Fernand Braudel, *The Structures of Everyday Life: Civilization & Capitalism, 15th–18th Century, Volume 1*, translated by Siân Reynolds (New York: Harper & Row, Publishers, 1979), 217. According to Braudel, as early as the sixteenth century, French cod fishermen in Olonne, near La Rochelle, engaged in a "race" every year to see which crew could make a fare to the Grand Bank and return the quickest. For more on the relationship between time management and modern, capitalistic, commercial enterprise, see E.P. Thompson, *Customs in Common: Studies in Traditional Popular Culture* (New York: The New Press, 1992), esp. chapter 6, "Time, Work-Discipline and Industrial Capitalism," 352–403. Thompson mistakenly argued that "fishing and seafaring peoples" maintained "a disregard for clock time," as opposed to urban industrial laborers. Ibid., 357.

[92] Schooner *Nancy*, 1795–1796, Ship's Log Books, microfilm 91, reel 3, JDPL. Emphasis my own.

it could be shipped to overseas markets. And, again, there was pressure to get the fish to market first.

It is possible, then, to precisely define the workers that labor recruiters such as Joshua Burnham typically hired, and we can closely detail the labor these workers performed at sea on the eve of the American Revolution. Young white men who lacked kinship ties to skippers or vessel owners dominated the industry. These maritime laborers habitually worked offshore banks. This work was very demanding. It took men such as Joshua Burnham and the crew of the *Polly* to deep-sea fishing waters in the spring, summer, and fall. The men worked long hours, far from home, and some never made it back to their families. The Atlantic Ocean claimed many workers' lives, and it broke many a limb. Fishermen persevered, however, and their labors built an important colonial industry.

Determining that most eighteenth century fishermen in colonial New England regularly worked deep-sea waters, especially those off the coast of Newfoundland, is key in terms of the origins of the American Revolution. Such deep-sea expeditions played an important role in the increased production capacity of the New England cod fisheries. Greater production meant greater profits for colonial fish merchants and an extended presence in the Atlantic economy. Having more fish to sell and making more profits from the cod trade also resulted in certain transatlantic commercial jealousies that contributed to the late eighteenth-century imperial crisis.

PART TWO

ATLANTIC ORIGINS OF THE AMERICAN REVOLUTIONARY WAR

Part Two explains why colonial fish merchants and fishermen decided to resist British authority during the late-eighteenth-century imperial crisis.

4

Cod and the Atlantic Economy

> The lumber and provisions of the American colonies were more necessary to the West India islands than the rum and sugar of the latter were to the former.[1]

While the crew scanned the Atlantic horizon for potential threats during the Seven Years' War, the twenty-nine-year-old master of a trade vessel fell asleep and dreamed of women.[2] The vessel carried a cargo of sugar products that had been received in exchange for fish. According to Ashley Bowen, "[O]ne hundred leagues at sea on my passage home from the West Indies in [the] sloop *Olive*, I dreamed I saw an elderly woman with a girl of about 10 years old a-standing by her side." After carefully considering the dream and the two females, Bowen concluded, "[T]he girl was the one I was to have." He believed he had a vision of his future wife's appearance in her youth, and she had a mole on her cheek. As soon as Bowen returned to his homeport of Marblehead, Massachusetts, the foremost fishing port in mainland North America, the superstitious mariner began searching for the woman of his dreams. Eventually, he met Dorothy Chadwick, a woman with a mole on her cheek, and in time they married. While such romantic dreams and quests for love were not common in records belonging to colonial New England mariners, the trade voyage to the West Indies and the exchange of fish for sugar had become a routine part of

[1] Adam Smith to William Eden, December 15, 1783. E.C. Mossner and Ian Simpson, eds., *Correspondence of Adam Smith* (Oxford University Press, 1977), 271.

[2] The dreams described in this paragraph can be found in *JAB*, I, 47.

New England's involvement in the Atlantic Economy during the eighteenth century. This had not been the case a century earlier.

The purpose of this chapter is twofold. It establishes that the volume of New England's Atlantic cod trade altered dramatically over the course of the seventeenth and eighteenth centuries. In addition, the direction of this overseas trade shifted between these eras. This trend and this shift fundamentally shaped political decisions on both sides of the Atlantic.

Resident New England commercial cod fisheries were born into the Atlantic world during the first half of the seventeenth century. Initially, plans for settlement either failed utterly or succeeded only in maintaining temporary, migratory commercial fishing operations.[3] It was not until 1631 that the first permanent commercial fishing station in the New World was established in Marblehead, Massachusetts.[4]

[3] According to Daniel Vickers, earlier attempts, including "Sagadahoc at the mouth of the Kennebec River (1607–1608), Wessagusset on the Massachusetts Bay (1622–1624), the Dorchester Company plantation on Cape Ann (1623–1624), and Mount Wollaston to the north of Plymouth (1623)," had all failed. Daniel Vickers, *Farmers and Fishermen: Two Centuries of Work in Essex County, Massachusetts, 1630–1830* (University of North Carolina Press, 1994), 90. For more on these unsuccessful attempts at settlement, see Todd Gray, "Fisheries to the East and West," in "The Distant-Water Fisheries of South West England in the Early Modern Period," in David J. Starkey, Chris Reid, and Neil Ashcroft, eds., *England's Sea Fisheries: The Commercial Sea Fisheries of England and Wales since 1300* (London: Chatham Publishing, 2000), 98–100; Faith Harrington, "'Wee Tooke Great Store of Cod-fish': Fishing Ships and First Settlements on the Coast of New England, 1600–1630," in Emerson W. Baker, *et al.*, eds., *American Beginnings: Exploration, Culture, and Cartography in the Land of Norumbega* (University of Nebraska Press, 1994), 191–216; Bernard Bailyn, *The New England Merchants in the Seventeenth Century*, 3rd ed. (Harvard University Press, 1995), 2–15; and Raymond McFarland, *A History of the New England Fisheries* (New York: D. Appleton & Company, 1911), 38–56. There was migratory fishing activity and some attempts at settlement at the Isles of Shoals during the early 1620s. John Scribner Jenness, *The Isle of Shoals: An Historical Sketch* (Boston: Houghton, Mifflin and Company, 1884), 46–57. However, there is no evidence of unbroken permanent settlement there. Also, Port Royal, Acadia, has been called "France's first permanent outpost in North America." Leslie Choquette, "Center and Periphery in French North America," in Christine Daniels and Michael V. Kennedy, eds., *Negotiated Empires: Centers and Peripheries in the Americas, 1500–1820* (New York: Routledge, 2002), 195. Port Royal, however, was abandoned between 1607 and 1610. When it was resettled, the port functioned as much or more as a fur trading entrepot and agricultural settlement than as a fishing station.

[4] Vickers, *Farmers and Fishermen*, 90–92; Christine Leigh Heyrman, *Commerce and Culture: The Maritime Communities of Colonial Massachusetts, 1690–1750* (New York: W.W. Norton & Company, 1984), 207; and Samuel Roads, Jr., *The History and Traditions of Marblehead*, 3rd ed. (Marblehead, MA: N. Allen Lindsay; Co., 1897), 8. Bernard Bailyn restricts his discussion to Salem, to which Marblehead was politically subservient

Why was the first permanent resident fishing station in the Western Hemisphere located in New England and not elsewhere? The New England region enjoyed certain comparative advantages over other regions. Its temperate climate and proximity to fertile lands stimulated the formation of a mixed agricultural/maritime economy. Warmer water temperatures yielded longer fishing seasons off the northeastern seaboard of North America as compared to the North Atlantic waters surrounding Newfoundland.[5] More northerly regions proved less conducive to permanent fishing settlements.

Terra do bacalhau, the "land of cod" in Portuguese, was a particularly harsh environment.[6] Peter Pope, a leading authority on early Newfoundland fishing, describes the island's seventeenth-century climate as "subarctic"; its soil as "glaciated... youthful and shallow"; and its ecosystem as "restricted" and "simple." He further notes that such "biogeographic instability" was responsible for the disappearance of two native populations prior to the 1600s.[7] Living conditions in Newfoundland were so challenging, in fact, that migratory English fishermen sent to temporarily live and work on the island often escaped to New England aboard trade ships. New England was a constant drain on Newfoundland's labor supply throughout the colonial period.[8]

until 1649. Bailyn, *The New England Merchants in the Seventeenth Century*, 17–18. Raymond McFarland wrongly attributes "the first permanent settlement" to "Pemaquid in 1625" and mistakenly puts the date of "the first fishing station" in Marblehead at 1633. McFarland, *A History of the New England Fisheries*, 51, 52–53. Peter Pope makes a strong archeological case that the Calvert experiment at Newfoundland between 1610 and 1629 did not fail utterly, as it left behind a certain amount of infrastructure for subsequent settlement. Peter E. Pope, *Fish into Wine: The Newfoundland Plantation in the Seventeenth Century* (University of North Carolina Press, 2004), esp. 45–78. Yet, even Pope is forced to acknowledge that Calvert and his servants decamped for warmer climes in Maryland. Moreover, the number of resident fishermen in Newfoundland did not outnumber the migratory fishermen until the 1790s. W. Gordon Handcock, *Soe longe as there comes noe women: Origins of English Settlement in Newfoundland* (St. Johns, NL: Breakwater Books, 1989), 74–85, esp. figure 4.2. In this chapter, a permanent resident fishing station is defined as an unbroken line of human settlement annually processing surplus cod for sale in overseas markets.

[5] Commercial fishing around Newfoundland was restricted to a short summer season. Gray, "Fisheries to the East and West," 98. For more on the longer fishing seasons off the New England coast, see Brian Fagan, *Fish on Friday: Feasting, Fasting and the Discovery of the New World* (New York: Basic Books, 2006), 260–261; Pope, *Fish into Wine*, 159; Daniel Vickers, *Farmers and Fishermen*, 116; and Harold Adams Innis, *The Codfisheries: The History of an International Economy* (Yale University Press, 1940), 80.

[6] Pope, *Fish into Wine*, 15.

[7] Ibid., 45–46.

[8] See the discussion in chapter six.

Farther south, early English settlers in colonial Virginia attempted to overcome their chronic food shortages through subsistence and commercial fishing in the Chesapeake Bay. Fish seemed to be most prevalent in the area during hurricane seasons, however. The warm climate also made spoilage an endemic problem and stymied commercial attempts to cure fish in the region.[9] As a result, therefore, of adverse climatic conditions to the north and south, the first permanent resident fishing station was established in New England.

It was not long before Marblehead had company. By the middle of the seventeenth century, permanent resident fishing stations had been established all along the New England coastline. There were stations at Pemaquid, Portland, and Falmouth in the Maine province; Richmond Island and Monhegan Island off the coast of Maine; at the Isles of Shoals off the coast of Maine and New Hampshire; at Portsmouth in New Hampshire; at Cape Anne, Salem, Marblehead, and Ipswich in Massachusetts; and at Nantasket (renamed Hull), Scituate, and Cape Cod in Plymouth.[10]

These resident fishing stations produced dried, salted cod, or saltfish, primarily for two major Atlantic markets. "Merchantable" grade saltfish was typically shipped to consumers in Italy, France, Spain, Portugal, and the Wine Islands. "Refuse" grades were shipped to sugar plantations in the West Indies.[11] Both of these markets were widely known by the seventeenth century. Thus, John Winthrop could write with confidence before departing England in 1630 that saltfish was "a known and staple Commodity."[12] Even the Puritan leader understood which markets demanded cod and why. These markets remained largely static throughout the colonial period. Europeans never demanded large amounts of refuse-grade cod, and sizeable quantities of valuable merchantable-grade cod were never shipped to slave plantation purchasers. The volume and

[9] James Wharton, *The Bounty of the Chesapeake: Fishing in Colonial Virginia*, 2nd ed. (University Press of Virginia, 1973), 3, 6–13, 27.

[10] Vickers, *Farmers and Fishermen*, 92–100; Heyrman, *Commerce and Culture*, 34, 209–210; Bailyn, *The New England Merchants in the Seventeenth Century*, 78; Innis, *The Codfisheries*, 75, 116; McFarland, *A History of the New England Fisheries*, 38–64. Gloucester turned toward farming after 1660 and did not come back to fishing in a systematic way until the first half of the eighteenth century. Plymouth was united with Massachusetts in 1691. The New England commercial fisheries were located primarily north of Cape Cod. Time and energy necessary for travel around the Cape seems to have restricted commercial fishing in colonial Rhode Island.

[11] The terms "merchantable" and "refuse" are defined in chapter two.

[12] *Winthrop Papers*, Vol. 2 (Boston: Massachusetts Historical Society, 1929), 145–146.

direction of the Atlantic cod trade, however, changed in significant ways between the seventeenth and eighteenth centuries.

The volume of dried, salted cod produced in colonial New England for export increased dramatically over the course of the seventeenth century. In 1641, John Winthrop recorded that 300,000 dried cod filets were produced in Massachusetts for distribution overseas.[13] Daniel Vickers estimates the weight of this yield at 6,000 quintals.[14] Given that one quintal represented 112 pounds in dry weight, the fish yield in Massachusetts in 1641 correlates to 672,000 pounds. In 1699, 50,000 quintals of dried cod were exported from New England.[15] Assuming that there were 54.5 fish in

[13] James Kendall Hosmer, ed., *Winthrop's Journal: "History of New England," 1630–1649*, Vol. 2 (New York: Barnes & Noble, Inc., reprint ed., 1959; orig. pub. 1908), 42.

[14] Vickers, *Farmers and Fishermen*, 99. Therefore, one quintal would have been the equivalent of fifty fish. For his part, Darrett B. Rutman puts the 1641 catch at "approximately 7,500 kentals." Darrett B. Rutman, "Governor Winthrop's Garden Crop: The Significance of Agriculture in the Early Commerce of Massachusetts Bay," *WMQ*, Vol. 20, No. 3 (July 1963), 403, footnote 25. This would mean that in 1641, one quintal was the equivalent of forty fish. Curiously, Vickers and Rutman both cite Winthrop's journal as the source of this data, but Winthrop does not provide a weight, only a total number of fish. See, Hosmer, ed., *Winthrop's Journal*, 42. Brian Fagan has recently estimated that one quintal was the equivalent of "about 120 dry fish." Fagan, *Fish on Friday*, 240n. It is possible that Fagan's data comes from a later period involving smaller average fish weights. This is not made clear. It is more likely that Fagan got his figure from Peter Pope, who used Richard Whitbourne's 1622 estimate that a quintal averaged "120 fish, dry." Pope, *Fish into Wine*, 33.

There does not seem to have been a standard correlation between fish counts and dry weight in eighteenth-century fish merchants' account books. This is not surprising, given the variation that existed in the size and weight of the cod. For example, in 1758, David Felt and the crew of the schooner *Molly* caught 34,400 fish, which, after processing, weighed 623.75 quintals. This correlates to fifty-five fish/quintal. One year later, on a first fare, Felt and crew caught 11,265 fish, which, after processing, weighed 215.25 quintals. This correlates to fifty-two fish/quintal. On the second fare of this year, 16,494 fish were caught, weighing 290 quintals, which correlates to fifty-seven fish/quintal. In 1760, Benjamin Henderson and the crew of the schooner *Esther* caught 23,689 fish on three fares, which, after processing, weighed 640.5 quintals. One quintal was, therefore, the equivalent of thirty-seven fish. One year later, on three fares, the same crew caught 22,718 fish, which, after processing, weighed 541.5 quintals. This correlates to forty-two fish/quintal. In 1762, the same crew caught 40,359 fish on just two fares, which, after processing, weighed 790 quintals. This correlates to fifty-one fish/quintal. In 1763, the same crew caught 53,164, which, after processing, weighed 757.5 quintals. This correlates to seventy fish/quintal. Timothy Orne Shipping Papers, JDPL. The average for these examples is 54.5 fish/quintal. Using this average would put the 1641 catch at 5,505 quintals.

[15] Innis, *The Codfisheries*, 118; and McFarland, *A History of the New England Fisheries*, 69.

a single quintal, then 2,725,000 fish were caught and processed in 1699.[16] This fish yield represents 5,600,000 pounds in dry weight. Between 1645 and 1675, the combined output of the New England fisheries expanded by five to six per cent annually, from 12,000 to 60,000 quintals per year.[17] In other words, 654,000 fish were caught in 1645, which weighed 1,344,000 pounds after processing, while 3,270,000 fish were caught in 1675, which weighed 6,720,000 pounds after processing.[18] Thus, the New England cod fishing industry produced five million pounds more of its leading export each year by 1675 than it had done thirty years earlier.

This production increase is primarily accounted for by New Englanders beginning to seize market share in the Atlantic cod trade at the expense of West Country fish interests. As a result of the English Civil War, West Country migratory fishing fleets at Newfoundland witnessed a decrease in output from 250,000 quintals in 1615 to 170,000 quintals in 1675.[19] This decrease in production was due to fewer and fewer English migratory fishing vessels heading westward. In 1605, the migratory fleet numbered some 250 vessels. By 1670, there were only eighty vessels in the transient fleet.[20] The number of English sack ships declined during this period as well. Sacks had been responsible for supplying the migratory fishing fleets that established temporary base camps in the region.[21] The Civil War and the three Anglo-Dutch Wars disrupted England's maritime economy in general and its commercial fishing industry in particular, as fishing vessels were converted to warships and fishermen were pressed into naval service. Bristol, the largest port in the West Country, became a battleground

[16] For the calculation of the average of 54.5, see footnote 14 above.

[17] Vickers, *Farmers and Fishermen*, 99–100. An economic analysis of early Essex County, the region with the highest concentration of fishing ports in North America in the seventeenth century, found "a 300%–500% increase in the size of the economy between 1640–1682." William I. Davisson, "Essex County Wealth Trends: Wealth and Economic Growth in 17th Century Massachusetts," *Essex Institute Historical Collections*, Vol. 103 (Salem, MA: Newcomb & Gauss, Co., 1967), 341.

[18] By comparison, 170,000 quintals were processed at Newfoundland in 1675. Pope, *Fish into Wine*, 425, table 37. This figure corresponds to 9,265,000 fish using the 54.5 fish/quintal average, which, after processing, would have weighed 19,040,000 pounds.

[19] David J. Starkey, "The Newfoundland Trade," in "The Distant-Water Fisheries of South West England in the Early Modern Period," in Starkey, Reid, and Ashcroft, eds., *England's Sea Fisheries*, 101–102.

[20] McFarland, *A History of the New England Fisheries*, 70.

[21] Pope, *Fish into Wine*, 104–112. Sizeable numbers of Dutch sack ships, freighters of 150–300 tons that purchased direct from the fishermen on the open sea, transported cod caught and dried at Newfoundland to European and Mediterranean markets from the 1620s. See Pope, *Fish into Wine*, 96–97, 99.

and changed hands twice, in 1642 and 1644, between Parliamentarians and Royalists. The West Country fleets that traveled to Newfoundland grew smaller and smaller and then stopped traveling to New England altogether.[22] European markets became hungry for new suppliers, and cod prices rose in several key European markets.[23] A subsequent section on the direction of New England's cod trade will show that New Englanders took advantage of the West Country's inability to supply Iberian markets with merchantable-grade cod. For now, it is enough to note that New Englanders capitalized on the West Country merchants' weakened position, increasing production in order to capture Iberian consumers' hearts and purses.

Over the course of the eighteenth century, colonial fish merchants witnessed sustained growth in their industry. The volume of dried, salted cod processed for export in New England continued to rise. In 1716, 120,000 quintals of this fishy commodity were processed in the colonies for overseas distribution. In 1731, this figure was 230,000 quintals. By 1765, the New England cod fishing industry was producing 350,000 quintals of dried, salted cod for export.[24] Put another way, 6,540,000 fish were caught and processed in 1716, which, after drying, weighed 13,440,000 pounds. By 1765, New England fishermen caught 19,075,000 fish, which, after processing, weighed 39,200,000 pounds. These figures correlate to an expansion in the annual production volume of 32,660,000 pounds between 1716 and 1765. To be sure, there were lulls in this pattern of increase, yet the overall trend in production over the course of the seventeenth and eighteenth centuries reflected expansion rather than contraction. To contemporaries, it must have seemed as though the potential for growth of New England's commercial fishing industry was virtually unlimited.[25] As subsequent chapters explain, however, various forces conspired against this colonial maritime expansion.

[22] Vickers, *Farmers and Fishermen*, 98; and Bailyn, *The New England Merchants in the Seventeenth Century*, 77–78. Also, see Pope, *Fish into Wine*, 132–44.

[23] Cod prices climbed by twenty-five per cent in Seville between 1637–1642 and 1644–1648 and rose fifty per cent in Madeira over the same periods, all in an effort to attract suppliers. See, Vickers, *Farmers and Fishermen*, 98.

[24] Vickers, *Farmers and Fishermen*, 154, table 4.

[25] Hence Adam Smith's 1776 remarks, mentioned in the introduction, that "the New England fishery in particular was, before the late disturbances, one of the most important, perhaps, in the world." Adam Smith, *An Inquiry into the Nature and Causes of the Wealth of Nations*, Max Lerner, ed. (New York: The Modern Library, 1937; orig., pub. in 1776), 544–545.

The volume of dried, salted cod produced in New England expanded over the course of most of the eighteenth century for several reasons. First and foremost, the expansion in the range of New England's deep-sea fishing operations during this period also provided Yankees with access to some of the best fishing waters the world has ever known.[26] Greater access to larger quantities and bigger sizes of cod contributed to the secular growth in colonial production during the eighteenth century. If colonial fish merchants had been unwilling or unable to invest capital and develop strategies for managing the greater risks associated with offshore commercial fishing, and if colonial maritime laborers did not risk life and limb in deeper waters, then it is likely that inshore waters would have been over-fished very quickly and production rates would have dwindled.

British victories in various colonial wars further stimulated production levels in the New England fishing industry at this time. The Treaty of Utrecht ended eleven years of war between France and Great Britain that had stunted the growth of the New England fishing industry. French warships and privateers had scoured the New England coast, slowing the traffic in cod considerably. Also, the peace treaty restricted French cod fisheries in the North Atlantic, thereby opening commercial possibilities to British interests, particularly in supplying the French West Indian islands with dried cod.[27] Likewise, British wars with France and Spain between 1743 and 1763 disrupted the Atlantic cod trade, yet the treaty signed at the end of the Seven Years' War, in 1763, transferred most of French Canada, including Newfoundland, into British hands.[28] The exclusion of French warships and most of their fishermen expanded New England's access to North Atlantic cod. One witness, most likely Brook Watson, a merchant in London with commercial ties to fish merchants in

[26] See the discussion in chapter three.

[27] Clive Parry, ed., *The Consolidated Treaty Series*, Vol. 27 (New York: Oceana Publications, Inc., 1969), 477–501.

[28] In article four of the treaty, France ceded Canada, Acadia, and Nova Scotia to Britain. Article five stipulated that "the subjects of France shall have the liberty of fishing and drying on a part of the coasts of the island of Newfoundland, such as is specified in the XIIIth Article of the Treaty of Utrecht [i.e. from Bonavista to Point Riche]." Article six of the treaty stipulated, "[T]he King of Great Britain cedes the Island of St. Pierre and Miquelon, in full right, to his Most Christian Majesty, to serve as a shelter for the French fishermen. And his said Most Christian Majesty engages not to fortify the said islands; to erect no buildings upon them, but merely for the convenience of the fishery; and to keep upon them a guard of fifty men only for the police." Parry, ed., *The Consolidated Treaty Series*, Vol. 42, 320–345. For copies of the treaty printed in colonial newspapers, see *Providence Gazette; And Country Journal*, May 21, 1763; *Newport Mercury*, May 23, 1763; and *New-Hampshire Gazette*, May 27, 1763.

New England, testified before Parliament in 1775 as to the 1763 treaty's positive effect on New England's cod trade to the West Indies:

> That the most inferior fish is exported to the neutral or French islands, and exchanged for molasses on VERY ADVANTAGEOUS TERMS, as the French are PROHIBITED from fishing. These molasses are sent to New-England, and manufactured into rum, which is sold for about fourteen pence per gallon, and used in the fisheries of New England and Newfoundland, and also exported to Guinea, and there exchanged for slaves, many of whom are sold to the French, and therefore eat the fish procured by the NEW ENGLANDERS.[29]

To be sure, the French continued to work North Atlantic waters, as the exclusionary articles in the peace treaties proved difficult to enforce. The British navy, however, forcibly removed French settlers and arrested French fishermen who violated these articles.[30] New Englanders, then, enjoyed greater access to cod and additional markets for this valuable commodity as a result of these treaties and their enforcement.

The volume of dried cod New England resident fishing stations produced in the eighteenth century did not automatically expand as quickly as the ink dried on these treaties, however. Certain merchants had to be convinced doing more in the Atlantic cod trade themselves was worth the additional effort, cost, and risk. In the case of Marblehead, Massachusetts, Reverend John Barnard, a fixture in the community from 1715 to 1770, became an outspoken advocate for locally-operated overseas trade. Barnard utilized his ecclesiastical position as pastor of Marblehead's First Congregational Church to bring about economic changes in the fishing industry. First, he traveled to Boston Harbor to meet with "English masters of vessels." Barnard educated himself on the finer points of transatlantic commerce – in his own words, "that I might by them be let into the mystery of the fish trade." After a time, Barnard felt that he had gained "a pretty thorough understanding" of what it meant to establish European connections, broker deals, and transport goods directly to a buyer instead of relying on middlemen in England or their colonial factors. "When I saw the advantages of it," Barnard wrote

[29] Colonial newspapers published the testimonies provided at these Parliamentary hearings for all to read or hear. *Massachusetts Spy: Or, American Oracle of Liberty*, May 3, 1775. The emphasized wording here can be found in the colonial newspaper, not the official Parliamentary records. For these records, see *AA*, Series 4, Vol. 1, 1643.

[30] There are numerous examples of the British navy arresting those French fishermen violating the tenets of the treaty in Colonial Secretary Letterbooks, 1752–1765, MUMHA. The navy further confiscated property and handed down sentences of banishment.

of transatlantic trade, "I thought it my duty to stir up my people, such as I thought would hearken to me, and were capable of practicing upon the advice, to send the fish to market themselves, that they might reap the benefit of it, to the enriching themselves, and serving the town." Once Barnard had acquired "a pretty thorough understanding of the business," the parson began disseminating his ideas about local economic development, especially among the younger men of his congregation.[31]

Joseph Swett Jr., for example, had been a merchant and a member of First Church. Barnard described him as "a young man of strict justice, great industry, enterprising genius, quick apprehension and firm resolution, but of small fortune." The pastor of First Church took the young Swett under his wing and informed him of his plan to establish transatlantic business connections to send local fish to overseas markets. "To [Swett] I opened myself fully," Barnard wrote, "laid the scheme clearly before him, and he hearkened unto me, and was wise enough to put it in practice." Soon, Swett fitted out a schooner, hired a captain, and "sent a small cargo [of fish] to Barbados." The venture proved so successful that Swett was able to outfit several trading vessels for additional voyages to Spain. In Barnard's words, Swett "soon found he increased his stock, built vessels, and sent fish to Europe, and prospered in the trade, to the enriching of himself; and some of his family, by carrying on the trade, have arrived at large estates."[32] According to Barnard, "the more promising young men of the town soon followed [Swett's] example." The result of the pastor's effort was that in 1766, Marblehead had "between thirty and forty ships, brigs, snows, and topsail schooners engaged in foreign trade."[33]

It is not surprising that, within the context of colonial New England, a single pastor had such an enormous impact on his community. A seventeenth-century Salem minister was responsible for generating public interest in the New England fisheries. The Reverend Hugh Peter urged Massachusetts's inhabitants to develop the local fisheries, and he further succeeded in securing overseas investment from merchants in England.[34] Moreover, whole English communities migrated overseas upon the insistent urgings of local pastors. The Reverend Ezekiel Rogers,

[31] John Barnard, "Autobiography of the Reverend John Barnard," *Collections of the Massachusetts Historical Society*, 3rd ser., Vol. 5 (Boston: Little, Brown, and Company, 1836), 239–40.

[32] Ibid.

[33] Ibid., 240.

[34] Robert G. Albion, William A. Baker, and Benjamin W. Labaree, *New England and the Sea* (Mystic, CT: Mystic Seaport Museum, Inc., 1972), 26.

for example, transplanted his East Riding farming community from England to the shores of the Merrimac River, where it contributed to the founding of a Puritan commonwealth during the mid-seventeenth century. In Rowley, Massachusetts, Rogers motivated his congregation to set up a cloth-making industry and negotiated "to break the Boston monopoly on imports by establishing a separate trading arrangement between his town and merchants in Yorkshire, [England]."[35] It was not unusual at all for ministers to play an active role in colonial economic affairs. Such advocacy contributed to the colonial fisheries' volumetric increases.

Local population booms in the eighteenth century also stimulated expansion in New England's production capacity.[36] More workers contributed to elevated productivity. New England fishing ports could send larger fleets to sea to catch and process greater volumes of dried, salted cod for export to overseas markets.

In addition to changes in the volume of dried, salted cod produced each year in colonial New England, the direction of the Atlantic cod trade shifted over time. The bulk of New England cod exports in the seventeenth century were shipped to Southern Europe. In 1641, John Winthrop reported that nearly all of the dried cod exported from Boston, New England's primary distribution center, went to Iberian ports.[37] In 1699, seventy-five per cent of the dried cod exported from Boston was shipped to Bilboa, Spain, alone.[38]

There are several explanations for the direction of New England's seventeenth-century Atlantic cod trade. Demand was high in Southern Europe, and elevated prices lured fish merchants to the region. Dried, salted cod was highly valued in this region for a variety of reasons. It was a cheaper alternative to fresh fish and meat. It conformed to Catholic dietary restrictions. It could stay dried for extended periods, and this durability made possible overseas distribution and additional transport inland from coastal ports. It proved exceptionally resilient to warm climates. Also, cod was high in protein.[39]

Iberian ports were also attractive markets for colonial fish merchants because of the amount of solid specie that they could obtain for their

[35] Stephen Foster, *The Long Argument: English Puritanism and the Shaping of New England Culture, 1570–1700* (University of North Carolina Press, 1991), 28–29.

[36] See the discussion in chapter three.

[37] Hosmer, ed., *Winthrop's Journal*, 42.

[38] Innis, *The Codfisheries*, 118; and McFarland, *A History of the New England Fisheries*, 69.

[39] This is discussed further in chapter one.

product. Indeed, the principal commercial function of the seventeenth-century colonial New England cod fisheries within the Atlantic economy was the transference of New World gold and silver into London coffers. Captain John Smith, that ubiquitous seventeenth-century English adventurer, famously stated: "Let not the meanness of the word fish distaste you, for it will afford us good gold as the mines of Guiana or Tumbata, with less hazard and charge, and more certainty and facility."[40] An English traveler who accompanied migratory fishermen to New England in 1623 observed numerous fishing vessels on the coast and commented: "[E]very one of them, by their confession, say that they have made good voyages, and now most of them are gone into Spain, to sell their fish where they have ready gold for it."[41] A seventeenth-century French traveler made a similar observation concerning cod fishing at Newfoundland: "[Y]ou make good Spanish coin out of it."[42] In the 1620s, one Parliamentarian noted that "the trade of fishing is a most beneficial trade to this realm for the increase of shipping, navigation, and mariners, and the bringing in of bullion and victual to a very great yearly value and supply."[43] "Fishermen," were "chiefly to be cherished for they bring in much wealth and carry out nothing."[44] Yet another member of Parliament maintained that, during the 1620s, the cod trade "brings in great store of bullion from Spain."[45] Still another Parliamentarian argued that the fishing industry was "a gain and surely returns money, *ex lege Hispania*."[46] The English government initially stipulated that Spanish merchants purchase, and English traders exchange, fish with gold and silver, instead of goods in kind. Such commercial regulation resulted from pressure by powerful business interests. According to one Parliamentarian, New England fish shipped to Spain "made no return but in treasure, for the Levant Company would not suffer them to deal in any commodities."[47] Specie restrictions were kept in place until 1661, when

[40] Cited in Wharton, *The Bounty of the Chesapeake*, 13.

[41] Quoted in Fagan, *Fish on Friday*, 269.

[42] Cited in Fernand Braudel, *The Structures of Everyday Life: Civilization & Capitalism, 15th–18th Century, Volume 1*, translated by Siân Reynolds (New York: Harper & Row, Publishers, 1979), 217–218. France maintained its own Atlantic cod fishery and strictly prohibited the importation of foreign cod.

[43] L.F. Stock, ed., *Proceedings and Debates of the British Parliament Respecting North America*, Vol. 1 (Washington, 1924), 70.

[44] Ibid., 60.

[45] Wallace Notestein, Frances Helen Relf, and Hartley Simpson, eds., *Commons Debates, 1621*, Vol. 4 (Yale University Press, 1935), 367.

[46] Ibid., Vol. 3, 81.

[47] Ibid., Vol. 4, 255–256.

Parliament changed its position and allowed goods and bills of exchange to be traded for fish.

The Price Revolution and increases in Catholic population levels further lured cod to Iberian ports.[48] The existence of large numbers of consumers and importers willing and able to pay high prices in solid specie meant that only the best grades of dried, salted cod were shipped to these markets. In 1642, *bacalao*, or *bacalhau*, fetched fourteen shillings per quintal in Spain and Portugal and eight shillings in the West Indies. For the period between 1662 and 1685, for which price records are particularly good, cod was worth twenty-seven to forty-four per cent more in Iberian ports than in West Indian ports.[49] In 1718, the British consul stationed at Lisbon, Portugal, reported that merchantable-grade, dried, salted cod sold for "at least 40 per Cent" more than refuse-grade cod.[50] At no point during the colonial period did Southern Europe lose this comparative price advantage.

There were other reasons for the strong commercial ties between New England and Southern Europe during the seventeenth century. New England cod's ability to be winter-cured gave it a small but important comparative advantage over Newfoundland cod, which was not typically dried on land during the winter months.[51] This winter cure meant that dried, salted cod from New England was somewhat fresher than Newfoundland fish for the spring Lenten season. Moreover, the slave

[48] Andalusia remained one of the major markets for dried Atlantic cod, and between 1500–1520 and 1595–1605 the price of cod rose 929% primarily as a result of increased demand. By comparison, over the course of the same time periods the general price index rose by 713%. See Earl J. Hamilton, "American Treasure and Andalusian Prices, 1503–1660: A Study in the Spanish Price Revolution," *Journal of Economic and Business History*, I (1928), 1–35, esp. 20–26; Innis, *The Codfisheries*, 51; and Ralph Davis, *The Rise of the Atlantic Economies* (Cornell University Press, 1973), 98–107. The influx of New World gold and silver helped consumers meet the inflated prices. According to Davis, following the fourteenth century Black Plague that reduced the population of Western Europe by one third, the population of Europe doubled between 1460 and 1620. Ibid., 16–17.

[49] Daniel Vickers, "'A Knowen and Staple Commoditie': Codfish Prices in Essex County, Massachusetts, 1640–1775," *Essex Institute, Historical Collections*, Vol. 124 (Salem, MA: Newcomb and Gauss, 1988), 198–202, table 1.

[50] William Poyntz to the Lords Commissioners of Trade and Plantations, Lisbon, September 17, 1718, NA SP 89/26/78A.

[51] Innis, *The Codfisheries*, 118; and McFarland, *A History of the New England Fisheries*, 69. Cod cured on land in the winter months tended to be fresher at the point when it was dried, as opposed to the warmer months of the year. Thus, New England cod sold in the spring was supposed to taste better than cod caught at Newfoundland over the summer and cured in the fall. Yet, due to Newfoundland's proximity to Europe, the price of Newfoundland merchantable cod caught in the summer tended to be a shade higher throughout the seventeenth century. See, Pope, *Fish into Wine*, 37–38, esp. table 2.

populations in the southern and West Indian colonies were not significant enough yet to attract the lion's share of New England's commercial attention. To be sure, culling practices had been established by this point, and refuse grades were being shipped south for sale to slave plantations in the southern mainland colonies and the West Indies.[52] Yet, the Southern European markets remained the dominant destination of New England cod during this initial phase.

During the eighteenth century, the direction of New England's Atlantic cod trade shifted. Colonial fish merchants shipped cod all over the Atlantic world at this time, but they placed the greatest emphasis on trade with the West Indies. The 1752–53 customs records for the port of Salem, Massachusetts, indicate only twenty-six per cent of the 121 vessels leaving Salem, Gloucester, and Marblehead cleared for Southern Europe and the Wine Islands in the year prior to the outbreak of the Seven Years' War that disrupted trade in the region. The remaining seventy-four per cent cleared for the Southern colonies and the West Indies. An overwhelming majority of the vessels from Salem and Gloucester cleared for the West Indies, while Marblehead sent roughly the same number to both destinations.[53] According to surviving customs records for 1764, Massachusetts vessels shipped 102,265 quintals of *merchantable*-grade cod at twelve shillings per quintal, for a total value of £61,359. At the same time, vessels from this colony shipped 137,794 quintals of *refuse*-grade cod at nine shillings per quintal, for a total value of £62,007.[54] Put another way, fifty-seven per cent of the dried, salted cod produced in Massachusetts in 1764 went to the West Indies. Between 1771 and 1773, New England fish merchants

[52] See the discussion of "merchantable" and "refuse" grades in chapter two.

[53] Heyrman, *Commerce and Culture*, 332n. According to David Hancock, the foremost authority on the role Madeira wine played in the Atlantic economy, Massachusetts only controlled twenty-five per cent of the market for dried cod in Madeira at any time. Moreover, between 1755 and 1774, the amount of dried cod cleared out of Salem and Boston for Madeira declined from 163,000 quintals to 122,000 quintals per year. David Hancock, "Markets, Merchants, and the Wider World of Boston Wine, 1700–1775," in Conrad Edick Wright and Katheryn P. Viens, eds., *Entrepreneurs: The Boston Business Community, 1700–1850* (Northeastern University Press, 1997), 85.

[54] William B. Weeden, *Economic and Social History of New England, 1620–1789*, Reprint ed., Vol. 2 (New York: Hillary House Publishers, Ltd., 1963), 750. Brook Watson, the London merchant with commercial ties to New England fish merchants who testified before Parliament in 1775, included "freight and all charges upon it at that time" in his calculation that the 102,265 quintals of merchantable grade cod was "valued at twenty Shillings per quintal," which was worth "the sum of £102,265." Watson also included freight rates and additional fees in his calculation that the 137,794 quintals of refuse cod was "valued at fourteen Shillings per quintal... and amounted to the sum of £96,455 16s." *AA*, Series 4, Vol. 1, 1638.

controlled eighty-two per cent of the West Indian market for British fish.[55] Other New England commodities followed this trade pattern. From 1768 to 1772, Yankee merchants shipped seventy-nine per cent of their grain and flour, eighty-eight per cent of their lumber, and ninety-nine per cent of their cattle and horses to the West Indies. In all, sixty-four per cent of New England's exports went to the West Indies during this time period.[56]

Qualitative evidence supports the quantitative material. Colonial fish merchants reported to peers who sat on Parliament's Treasury Board in 1750 that "one half" of the dried, salted cod caught and processed in New England went to the West Indies to be exchanged for sugar products.[57] In 1763, as news of international peace spread around the Atlantic world, Edward Payne, a Boston merchant and member of the Society for the Encouragement of Trade and Commerce, estimated that sixty per cent of New England's annual catch was being shipped to sugar plantations in the Caribbean.[58] One year later, Robert Hooper and Jeremiah Lee, two of the wealthiest fish exporters in Marblehead, informed Lieutenant Governor Thomas Hutchinson that "the fish caught in the several fishing Towns in the Province of the Massachusetts Bay is annually from 220 to 240 Thousand Quintals, about 2/5 whereof is made merchantable & shipped to Spain, Portugal, & Italy; the other 3/5 is shipped for the West Indies, not being fit for the European Markets."[59] The figure of three-fifths was later corroborated in 1766 by "Mr. Kelly," who "lived in New York and was bred a merchant" and two merchants whose "knowledge is principally confined to New England, Nova Scotia and

[55] Selwyn H.H. Carrington, *The British West Indies During the American Revolution* (Providence, RI: Foris Publications, 1988), 44, Table 30. I arrived at the figure of eighty-two per cent by first adding the barrels, quintals, and kegs of fish imported into the West Indies between 1771–73 from America, Nova Scotia and Canada, and Newfoundland. I then divided the sub-total for America by the total. Here, I assume that "America" refers to the New England cod fisheries. In 1771 alone, New England fish merchants shipped ninety-two per cent of all fish exported from British North America to the "British & Foreign West Indies." "Exports to the British & Foreign West Indies between the 5th day of January 1771 & the 5th day of January 1772," British North American Customs Papers, 1765–1774, MHS.

[56] Stephen J. Hornsby, *British Atlantic, American Frontier: Spaces of Power in Early Modern British America* (University Press of New England, 2005), 135–136, figure 4.6.

[57] Cited in Weeden, *Economic and Social History of New England*, Vol. 2, 641.

[58] Testimony of Edward Payne of Boston, Merchant, Boston, December 1763, Ezekiel Price Papers, 1754–1785, MHS.

[59] Hooper Fisheries Statement, 10 January 1764, MHS, Misc. Bd. Manuscripts, 1761–1765.

Newfoundland": "Mr. Wentworth" and "Brook Watson."[60] In addition, Benjamin Pickman and Samuel Gardner, fish merchants in Salem, Massachusetts, calculated "that there was Thirty Sail of Vessels Employed on the Bank Fishery in the year 1762, and that the Quantity of Fish taken by them that Season was Twenty six Thousand Seven hundred & Forty Quintals, whereof Six Thousand Two hundred & Twenty Three Quintals was Merchantable, Twenty Thousand Five hundred & Seventeen Refuse."[61] Put another way, Pickman and Gardner determined that twenty-three per cent of Salem's annual catch went to Europe, while seventy-seven per cent went to the West Indies.

In his capacity as Secretary of State under President George Washington, Thomas Jefferson retrospectively calculated that a nearly equal amount of trade went to the West Indies and Southern Europe between 1765 and 1775. In Jefferson's report, fifty-one per cent of New England's dried cod was being exported to Europe, which in this case refers specifically to the ports in Iberia and the Wine Islands, while forty-nine per cent of the cod was exported to the West Indies.[62] There is reason, however, to suspect that Jefferson may have overestimated the trade to Europe, given the modern analysis that has been done on customs records, Hooper and Lee's report, and the Parliamentary testimonies of three additional merchants. Taken together, the quantitative and qualitative trade data indicates eighteenth century New Englanders placed a greater emphasis on shipping dried cod to West Indian slave plantations than they did during the seventeenth century.[63] Indeed, such was the commercial connection between Massachusetts and the West Indies that the region southwest of Boston became known as "Jamaica Plain," and a body of water there was called "Jamaica Pond."[64]

There are several explanations for the shift in direction of New England's Atlantic cod trade in the eighteenth century. The nature of the

[60] *PDBP*, Vol. 2, 359–365, 571–572. This is almost certainly the same Brook Watson who testified before Parliament on behalf of the colonial fishing industry later in 1775.

[61] Benjamin Pickman and Samuel Gardner of Salem, to Edward Payne and Thomas Gray in Boston, Salem, January 10, 1764, Ezekiel Price Papers, 1754–1785, MHS.

[62] Thomas Jefferson, Secretary of State, "Report on the State of the Cod Fisheries, 1791," *American State Papers: Commerce and Navigation*, I:13.

[63] The evidence here contradicts certain modern scholarly interpretations. James G. Lydon estimates, "[T]he West Indies and southern Europe were of nearly equal importance as market areas" in the eighteenth century. James G. Lydon, "Fish for Gold: The Massachusetts Fish Trade with Iberia, 1700–1773," *NEQ*, Vol. 54, No. 4 (December 1981), 539. For more of the same, see Innis, *The Codfisheries*, 162.

[64] "Plan of Boston and vicinity, 1775," *NDAR*, Vol. 1, 197.

curing technique New Englanders used changed over time. The proliferation of distilleries in eighteenth-century New England and colonists' expanding taste preference for sweet things necessitated increased sugar imports. Increased West Indian slave populations required additional provisions. The Newfoundland interests' domination of the Southern European markets effectively crowded out New Englanders. Markets with slaves also proved very reliable. Each of these factors will now be detailed in turn.

The curing techniques used in the fishing industry to convert catches into commodities influenced the direction of New England's Atlantic cod trade. Before the development of an offshore fishery, New England fishermen caught fish and headed for land as fast as possible, where they dressed, salted, and spread cod on fish flakes for drying.[65] This process usually resulted in high yields of merchantable-grade, dried cod capable of being marketed in Europe. Because of the distance involved, fishing on the banks required a different curing process. New Englanders began utilizing the two-part combination cure first developed by the Dutch in the late fourteenth century.[66] Schooners traveled several days to offshore banks, where they took in cod, and fishermen salted the catch wet, right out of the ocean. This salting was the first part of the cure. Men then stacked the lightly salted cod in the holds of their schooners, where the fish sat for prolonged durations. Later, the crew returned to their homeport, where shoremen air-dried the damp, salty catch. The air-drying completed the cure. Invariably, the combination of a wet-salt cure and an air-dry cure produced a greater percentage of refuse-grade, dried cod. Indeed, one New England fish merchant testified before Parliament in 1766 that colonial fishermen worked "at a great distance from the shore," and their catch was "kept in ships where it grows soft and becomes therefore soft and refuse."[67] Such "refuse" could only be sold to slave plantations. In Payne's words, "as we can't cure Fish for the European Market separate from the other sort sent to the W. Indies, & as we have no other Market

[65] Vickers, *Farmers and Fishermen*, 149.

[66] See the discussion in chapter three.

[67] *PDBP*, Vol. 2, 364. Similarly, William Poyntz, British consul at Lisbon, Portugal, observed in 1718, "As to the New England fish, it is often taken at that Distance from the Land that it is several Days ere they get it in; & though they carry Salt to Sea to salt it, yet it lies so long in the Boat unwashed makes it very unfit to dry as they do (to impose on the ignorant buyers)." William Poyntz to the Lords Commissioners of Trade and Plantations, Lisbon, September 17, 1718, NA SP 89/26/78A. Poyntz believed that the cod New Englanders caught "ought to be cured for wet saltfish," as opposed to using the combination cure to get a dry, salted product.

for what is made by the Bankers [i.e. the refuse-grade, dried cod], it will be lost if not sent to the foreign Islands in the W. Indies." It was Payne's considered opinion that without West Indian markets, and the slave populations therein, "the whole Bank Fishery" would have been "destroy[ed]."[68]

The proliferation of New England rum distilleries in the eighteenth century provided additional stimulus for the eighteenth-century shift in Atlantic trade patterns. There were a handful of distilleries in operation throughout New England at the turn of the eighteenth century. But, by 1770 there were 140 rum distilleries in the thirteen North American colonies, ninety-seven of which were located in New England. At this time, colonists annually imported over six-and-a-half million gallons of molasses from the West Indies. Lumber and dried cod were the two primary commodities New Englanders exchanged for the enormous quantity of molasses they needed to feed their expanding rum distilleries, and this molasses could not be obtained from Iberia or the Wine Islands.[69] The West Indies were the key suppliers.

Maritime laborers involved in the New England fishing industry were also great consumers of West Indian sugar products.[70] According to one expert witness called to testify before Parliament in 1766, "the poorer sort of people in North America," which included a majority of those who worked in the cod fishing industry, "use the [imported West Indian] molasses in making small beer."[71] The same witness testified that the molasses was also "distilled into rum which is employed in carrying on the fishery and the Guinea trade."[72]

An above average social value was typically assigned to drinking within maritime cultural circles.[73] Commercial fishermen, in particular, had a

[68] "State of the Trade & Observations on the Late Revenue Acts," 1767, Ezekiel Price Papers, 1754–1785, MHS.

[69] John J. McCusker and Russell R. Menard, *The Economy of British America, 1607–1789* (University of North Carolina Press, 1985), 290–291, esp. figure 13.1; Weeden, *Economic and Social History of New England*, Vol. 2, 502, 641; and Richard Pares, *Yankees and Creoles: The Trade between North American and the West Indies before the American Revolution* (London: Longmans, Green and Co., 1956), 31–33.

[70] Anglo-Americans first developed taste preferences for sweet things in the 1700s. Sidney W. Mintz, *Sweetness and Power: The Place of Sugar in Modern History* (New York: Viking Press, 1985), 18.

[71] *PDBP*, Vol. 2, 362.

[72] *PDBP*, Vol. 2, 360. The same colonial merchant later testified "that the fisheries would take off more rum than the Guinea trade, but that both will take off great quantity." Ibid., 364.

[73] See the discussion in Christopher Magra, "Faith At Sea: Exploring Maritime Religiosity in the Eighteenth Century," *IJMH*, Vol. 19, No. 1 (June 2007), 93.

long and wide reputation for loving their alcohol.[74] On average, British North American colonists consumed 8.5 million gallons of rum each year in the early 1770s.[75] This amounts to 3.6 gallons of annual per capita consumption.[76] In Massachusetts in 1763, fishermen typically consumed over seven gallons of rum each year, almost twice the colonial average for the 1770s.[77] Henry Florence Jr., a Marblehead fisherman, died while working on the fishing schooner *Philip* in 1771 "as a result of a dare to see who could consume the most New England rum."[78] This rum was imported directly from the West Indies, and it was distilled from imported molasses in the colonies.

New Englanders both consumed the rum, and they shipped it to Newfoundland to supply fishermen there.[79] Salem fish merchants calculated

[74] See the commentary of John Josselyn, an English traveler who visited the New England coast during the 1630s and then again in the 1660s, in Lindholdt, ed., *John Josselyn, Colonial Traveler*, 144–45; and that of Commodore Hugh Palliser, acting governor at Newfoundland, Palliser to Justice M. Gill, St. Johns, October 2, 1764, Colonial Secretary Letterbooks, 1752–1765, MUMHA.

[75] McCusker and Menard, *The Economy of British America*, 290.

[76] The total population used in this calculation is 2.35 million. For this figure, see Alice Hanson Jones, *Wealth of a Nation To Be: The American Colonies on the Eve of the Revolution* (Columbia University Press, 1980), 37, table 2.2.

[77] In 1763, Salem fish merchants calculated that workers in the colonial Massachusetts's fishing industry consumed "600 hogsheads of rum" during the fishing season (i.e., spring, summer, and fall). Benjamin Pickman and Samuel Gardner of Salem to John Rowe, Esq., and others Committee of the Society for Encouraging Trade & Commerce located in Boston, Salem, December 24, 1763, Ezekiel Price Papers, 1754–1785, MHS. In the eighteenth-century British Atlantic world, there were 52.5 gallons in a hogshead. Roger Morriss, "The Supply of Casks and Staves to the Royal Navy, 1770–1815," *Mariner's Mirror*, Vol. 93, No. 1 (February 2007), 50, footnote 8. Thus, workers in Massachusetts's fishing industry consumed a total of 31,500 gallons of rum each year. Taking into account the fact that 4,405 workers labored in this colony's fisheries, each worker consumed over seven gallons of rum each year. For the size of the workforce, see Thomas Jefferson, Secretary of State, Report on the State of the Cod Fisheries, 1791, *American State Papers: Commerce and Navigation*, I:13, table 2. According to Daniel Vickers, "fishing crews departing on four-to-eight-week voyages normally carried about 12 gallons of rum and at least 60 gallons of cider (Imperial measure). Among a crew of seven or eight members (including two to four boys), this would amount to roughly a quart of cider and six ounces of rum per fishermen per day." Vickers, *Farmers and Fishermen*, 182, footnote 61. Vickers admits, however, that "this must be regarded as a minimal estimate since crew members sometimes purchased more in ports closer to the grounds."

[78] *JAB*, II, 652.

[79] In 1763 alone, 49,140 gallons of rum were shipped from Boston to Newfoundland. This figure represents eighty-six per cent of all rum exported from the colonies to Newfoundland. See, Head, *Eighteenth Century Newfoundland*, 120, table 6.10. Ralph Greenlee Lounsbury dates the genesis of New England's rum shipments to Newfoundland at 1675. Ralph Greenlee Lounsbury, *The British Fishery at Newfoundland 1634–1763*

in 1763 that "Boston alone" supplied "Nova Scotia & Newfoundland" with "993 hogsheads of rum and 301 hogsheads of molasses."[80] A writer concerned with the Newfoundland fishery explained in 1807 that without access to "strong drink" it "would be impossible to continue the [fishing] trade [i.e. industry], for ten hours in the boats every day in the summer and the intolerable cold of the winter makes living hard."[81] New England was thus an importer and an exporter of an alcoholic beverage that was said to ease the plight of the working man. Rum could not have been produced without molasses, a by-product of sugar production. Rum, small beer, and molasses go a long way toward explaining the increased commercial ties between New England and the West Indies.[82]

There was also greater demand in the West Indies for New England's dried, salted cod during the eighteenth century. The plantation complexes in the West Indian islands imported huge numbers of black African slaves at this time. Jamaica became the largest slave society in the British West Indies, expanding its slave population twenty-five times over between 1673 and 1774.[83] In 1700, there were 115,000 slaves in the British West Indian islands and 54,000 in the French islands. Forty years later, these numbers expanded to 250,000 and 228,000, respectively. By 1770, there were 434,000 slaves living, working, and dying on the British islands and 393,000 in the French islands.[84] In short, British planters had to

(Yale University Press, 1934; reprinted New York: Archon, 1969), 191–192. However, Pope, Head, and Bailyn maintain that Boston merchants were trading general supplies, of which rum may have been involved, as early as 1640. Pope, *Fish into Wine*, 151; Head, *Eighteenth Century Newfoundland*, 111; and Bailyn, *The New England Merchants in the Seventeenth Century*, 129.

[80] Benjamin Pickman and Samuel Gardner of Salem to John Rowe, Esq., and others, Committee of the Society for Encouraging Trade & Commerce located in Boston, Salem, December 24, 1763, Ezekiel Price Papers, 1754–1785, MHS.

[81] Quote taken from Innis, *The Codfisheries*, 103. According to Innis, as a result of fishermen's great consumption of alcohol, there were seventy-four taverns in Newfoundland in 1723, fifty in St. John's alone. The figure for Newfoundland increased to 122 in 1750. Ibid., 153–154. For an excellent discussion of seventeenth-century fishermen and their consumption of alcohol, see Pope, *Fish into Wine*, 393–401.

[82] For more on colonial rum production and exports of molasses from the West Indies to North America, see John J. McCusker, *Rum and the American Revolution: The Rum Trade and the Balance of Payments of the Thirteen Continental Colonies*, 2 Vols. (New York: Garland Publishing, Inc., 1989). Also, see David Eltis, "New Estimates of Exports from Barbados and Jamaica, 1665–1701," *WMQ*, Vol. 52, No. 4. (October 1995), 631–648.

[83] Andrew Jackson O'Shaughnessy, *An Empire Divided: The American Revolution and the British Caribbean* (University of Pennsylvania Press, 2000), 8, 27.

[84] Stanley L. Engerman, "Economic growth of colonial North America," in John J. McCusker and Kenneth Morgan, eds., *The Early Modern Atlantic Economy* (Cambridge

feed 319,000 more mouths in 1770 than they did at the beginning of the century, while French planters had 339,000 more.

Feeding these slaves well, and providing for their living conditions in general, were not issues plantation managers and owners spent a great deal of time worrying about in the West Indies. To be sure, black slaves did not live or eat as well as their white masters, yet slaves remained consumers of food nonetheless. They produced some of the protein for their dietary needs. Slaves planted and grew their own vegetables in gardens and on the edges of cultivated sugar plantations. A few raised poultry instead of planting food. Plantation owners also provided slaves with fishing hooks to help supplement the workers' diets, and they became proficient fishermen. Such cost-saving measures, however, were not enough to make these coerced laborers self-sufficient. Slave provision grounds tended to be small, most of the fresh fish that was caught supplied masters' tables, and hurricanes frequently devastated food crops on the islands. As a result, the majority of West Indian slave diets were comprised of imported provisions. Some salted beef was brought in from Ireland, although most of it went to fuel the remaining white indentured servant populations on the islands. Fish merchants in England, Scotland, and the Netherlands shipped dried and pickled herring to the West Indies, and a portion went to the slaves. North America provided a variety of provisions, including flour, rice, corn, and beans, yet dried, salted, refuse-grade cod from New England remained an inexpensive source of protein with which to fuel slave labor.[85]

University Press, 2000), 236, table 9.1A. The table lists only "black" populations. Here, it is assumed a majority of this West Indian black population were slaves. According to O'Shaughnessy, "free people of color represented less than 2 per cent of the total population of the islands" during the mid-to-late eighteenth century. O'Shaughnessy, *An Empire Divided*, 29. For more on the expansion of the French West Indian islands, see W.J. Eccles, *The French in North America, 1500–1783*, revised edition, (Michigan State University Press, 1998), chapter 6, "The Slave Colonies, 1683–1748"; and Robin Blackburn, *The Making of New World Slavery: From the Baroque to the Modern, 1492–1800* (New York: Verso, 1997), 277–306, 431–456.

[85] Sherry Johnson, "El Niño, Environmental Crisis, and the Emergence of Alternative Markets in the Hispanic Caribbean, 1760s70s," *WMQ*, 3rd Series, Vol. LXII, No. 3 (July 2005), 365–410; J.R. Ward, *British West Indian Slavery, 1750–1834* (Oxford: Clarendon Press, 1988), 19–24; R.C. Nash, "Irish Atlantic Trade in the Seventeenth and Eighteenth Centuries," *WMQ*, 3rd Series, Vol. 42, No. 3 (July 1985), 334–336; Richard N. Bean, "Food Imports into the British West Indies: 1680–1845," in Vera Rubin and Arthur Tuden, eds., *Comparative Perspectives on Slavery in New World Plantation Societies* (New York: New York Academy of Sciences, 1977), 581–590; Carrington, *The British West Indies During the American Revolution*, 53–56, 110–114; Richard B. Sheridan, "The Crisis of Slave Subsistence in the British West Indies during and after

George Walker, an absentee sugar planter from Barbados and a colonial agent for the island, testified before Parliament on March 16, 1775, as to the island's dependence on imported provisions. He stated:

> The land and labor of the country being devoted to the cultivation of the Sugar-cane, the Corn and Provisions they raise are merely accidental; they are no more than can be raised without prejudice to the Sugar Cane. To the Sugarcane every thing is sacrificed as a trifle to the principal object. In Barbados, I doubt whether the Corn, (it is Indian Corn, not Wheat) and the ground Provisions (I mean Yams and other Roots,) raised in the Island, are sufficient to maintain the inhabitants for three months; I am certain they will not maintain them for four months, unless the four months be those in the beginning of the year, in the season for ground Provisions. The Indian Corn and ground Provisions cannot, by common means, be preserved for any length of time.... dry weather, or excess of wet weather, hurricanes, blast, vermin, frequently diminish or destroy the hopes of the Planter... As to the Leeward Islands, they produce neither Corn nor ground Provisions worth mentioning, except Tortola. Tortola was a Cotton Colony; Cotton and Corn are not inconsistent.... it follows, that such Colonies must depend, in proportion to the extensiveness of the manufacture [i.e. sugar production], upon other places for necessary food, for actual subsistence. The observation applies to Jamaica, and to the Islands under the Granada Government.[86]

the American Revolution," *WMQ*, 3rd Series, Vol. 33, No. 4 (October 1976), 620–621; Pares, *Yankees and Creoles*, 24–27; Richard Price, "Caribbean Fishing and Fishermen: A Historical Sketch," *American Anthropologist*, Vol. 68, No. 6 (December 1966), 1363–1383, esp. 1370; Eccles, *The French in North America, 1500–1783*, 90, 193–194; and Innis, *The Codfisheries*, 162–163. Blackburn maintains that the French islands were less dependent on imported provisions than the British islands. Blackburn, *The Making of New World Slavery*, 438–439. O'Shaughnessy argues that Barbados and the British Leeward Islands were more dependent on external food supplies than Jamaica and the British Windward Islands. O'Shaughnessy, *An Empire Divided*, 70–72. For more on the importance of slave provision-grounds to Jamaica's domestic economy, see Beth Fowkes Tobin, "'And there raise yams': Slaves' Gardens in the Writings of West Indian Plantocrats," *Eighteenth-Century Life*, Vol. 23, No. 2 (May 1999), 164–176; Sydney W. Mintz and Douglas Hall, "The Origins of the Jamaican Internal Marketing System," in Hilary Beckles and Verene Sheperd, eds., *Caribbean Slave Society and Economy* (New York: New Press, 1991), 319–334; Orlando Patterson, *The Sociology of Slavery: An Analysis of the Origins, Development and Structure of Negro Slave Society in Jamaica* (London: MacGibbon & Kee, 1967), 216–230; and John H. Parry, "Plantation and Provision Ground: An Historical Sketch of the Introduction of Food Crops into Jamaica," *Revista de historia de America*, Vol. 39 (1955), 1–20. More comparative work needs to be done to determine the exact extent to which British and French islands depended on supplies of New England dried cod.

[86] *AA*, Series 4, Vol. 1, 1722–1723.

Walker further testified that North America, Ireland, and England sent provisions to the sugar islands, but the colonial agent and planter stressed the islands' utter dependence on North America.

John Ellis, an absentee sugar planter from Jamaica, testified before Parliament at the same time as Walker. Ellis reinforced Walker's position that the British West Indies depended on North American provisions. The Jamaican planter stated that land on Britain's most valuable sugar island had been "appropriated to the purposes of raising Cattle." Additional farmlands "afford a partial supply of Provisions, such as Plantains, Roots, and Indian Corn." Nevertheless, "on the whole," he concluded, "though the Island of Jamaica has, in respect of internal supplies, greatly the advantage over Barbados and the other British Sugar Islands [in terms of domestic provisions], yet from the circumstances of drought and gusts of wind which happen frequently, and are particularly destructive to the Plantain Trees, which yield the chief support of the Negroes, her dependence on North America in point of Provisions, is very great."[87]

Dried, salted cod played a particularly important role in provisioning these West Indian slaves. According to Barbados's colonial agent, "Salted Fish . . . is the meat of all the lower ranks of people in Barbados, and the Leeward Islands. It is the meat of all the Slaves in all the West Indies."[88] Ellis similarly observed that Jamaica "receives from America great quantities of salted Fish; which, with Herrings from Europe, serve the Negroes as Meat."[89] On average, a single slave living and working on the British islands consumed thirty pounds of fish on an annual basis during the second half of the eighteenth century. This figure varied depending on the availability of provisions and the plantation involved. Slaves could each be given as little as twelve pounds of fish per year or as much as seventy pounds. The most common daily allowance for a slave at this time was one pint of corn and one-seventh of a pound of fish.[90] Thus, increased demand for provisions generated greater commercial ties between New England and the West Indies.

Newfoundland fishing interests regained control over Iberian markets after their decline during the mid-seventeenth century, and this further influenced the direction of New England's Atlantic cod trade. England had recovered from its Civil War and expanded its naval power and its

[87] Ibid., 1731.
[88] Ibid., 1723.
[89] Ibid., 1732.
[90] Ward, *British West Indian Slavery*, 20–21, 105–107, esp. table 7.

commercial presence at Newfoundland. Resilient West Country migratory fishermen began working the deeper waters of the Grand Bank off the coast of the land of cod around the turn of the eighteenth century. These bankers, along with bye-boat fishermen, men who left their vessels moored at Newfoundland only to hitch a ride back as passengers the following season, increased English production from 125,000 quintals in 1684 to 347,000 quintals in 1692.[91] Greater production helped the Newfoundland interests to control an average of seventy-nine per cent of all Southern European markets in Spain, Portugal, the Madeiras, Azores, and Canaries throughout the entire eighteenth century.[92] By contrast, these same Newfoundland interests controlled only fourteen per cent of the West Indian markets between 1771 and 1773.[93] The proximity of Newfoundland and West Country exporters to Southern European markets gave them an edge over New Englanders in terms of the time that it took to get their fish to consumers. Lower transportation costs meant that the Newfoundland interests could charge less for their product in Southern Europe than the New Englanders, which helped them gain market dominance in the region. This economic reality provided additional incentive for New Englanders to concentrate their commercial efforts on the West Indies.

One other factor may have influenced the eighteenth-century shift in the direction of New England's Atlantic cod trade. There was a seemingly constant demand for lesser grade fish among the slave plantations throughout the year. Steady demand would have meant relatively stable price levels. Such price reliability would have also meant that fish merchants could be surer of their profit margins in trade deals with these plantations. Gloucester fish merchant Daniel Rogers recorded in his account book for 1770 the prices of merchantable-grade, dried cod over the course of the year. He documented that the spring, or first, fare sold for eighteen shillings per quintal. The summer, or second, fare sold for fifteen shillings per quintal. The fall, or third, fare sold for sixteen shillings per quintal. Assuming 100 quintals of merchantable fish were sold after each catch had been processed, then, all things being equal, the spring fare fish at the above price would fetch £90; the second fare fish would fetch £75; and the third fare fish would fetch £80. Thus, the price of merchantable-grade, dried, salted cod could drop seventeen per cent in a given year, presumably due to the end of the Lenten season and declining demand.

[91] See Starkey, "The Newfoundland Trade," 102–103.
[92] Lydon, "Fish for Gold," 544–545, table 2.
[93] Carrington, *The British West Indies during the American Revolution*, 44, table 30.

The price trend of merchantable fish was, therefore, to drop over the course of the year. By contrast, refuse-grade, dried, salted cod tended to maintain its price buoyancy over the same duration, shifting only during the fall fare from £0.13.4/quintal to £0.14.0/quintal – a mere five per cent fluctuation.[94] If such price trends held constant for all colonial fish merchants over an extended period of time, then the plantation markets would have been less lucrative but more stable. Such price reliability would have reduced risks associated with this trade and made the West Indies that much more attractive.

For all of these reasons, eighteenth-century maritime laborers such as Ashley Bowen routinely sailed between New England and the West Indies. Those who worked in the fishing industry commonly fished during the spring, summer, and fall. They then worked for their fish merchant employers during the winter months, often on the same vessels, on trade voyages distributing refuse-grade, dried, salted cod to purchasers for slave plantations primarily in the West Indies. As for the *Polly's* crew mentioned in the last chapter, Joshua Burnham worked for the Choats as skipper/master of various Ipswich schooners on trips to offshore banks and then to ports in the Southern and West Indian colonies throughout the 1760s and early 1770s. For a period of nine years between 1766 and 1774, John Andrews labored as sharesman/mate alongside Burnham on these expeditions/voyages.[95] The Ipswich crew of the schooner *Neptune* worked on fishing fares in the spring, summer, and fall seasons between 1768 and 1770 before sailing south each winter to haul refuse-grade, dried, salted cod to Dominica in the West Indies.[96] Gloucester laborers worked on board the schooner *Liberty* on twelve fishing fares between 1770 and 1773. During the winter months, these men set sail and transported fish south to undisclosed locations in the West Indies.[97] Following fishing fares, Salem fish merchant Timothy Orne's employees on board the schooner *Molly* sailed to Monte Christo in the West Indies in 1760 and 1761.[98] The crew of a Marblehead schooner by the same name made regular winter trips to Barbados between 1766 and 1771 after their fishing fares.[99] Seth Harding commanded several schooners out of Cape

[94] Daniel Rogers Account Book, 1770–1790, JDPL.
[95] Joshua Burnham Papers, 1758–1817, JDPL.
[96] Joshua Burnham Papers, 1758–1817, Schooner *Neptune*, [1766–1770], box 1, folder 3, JDPL.
[97] Daniel Rogers Account Book, 1770–1790, JDPL.
[98] Timothy Orne Shipping Papers: schooner *Molly*, 1758–60, box 7, folder 11; and schooner *Molly*, 1761–66, box 7, file 12, JDPL.
[99] Richard Pedrick Account Book, 1767–1784, MDHS; and Richard Pedrick Papers, schooner *Molly*, [c. 1766–71], box 2, folder 28, MDHS.

Cod on trade voyages to the West Indies during the mid-eighteenth century, most likely hauling refuse cod, before moving to Nova Scotia to establish a salmon fishery.[100] For all of these maritime laborers, winter trade voyages to the West Indies made annual earnings from fish merchant employers possible, and they closely tied colonial New England to the West Indies.

In sum, the volume and direction of New England's Atlantic cod trade changed in dramatic ways over the course of the colonial period. Colonial producers realized a protracted increase in the amount of dried, salted cod available for export. The production, distribution, and consumption of this valuable commodity contributed to greater commercial ties between colonial New England and the West Indies in the eighteenth century than was previously the case. This shift was largely the result of a combination of Newfoundland's control over the Southern European markets, New Englanders' choice of curing techniques, the proliferation of distilleries and colonists' taste for sweet things, the expansion of West Indian slave populations, and the reliability of markets with slaves. As a result of these factors, North-to-South trade routes between New England and the West Indies came to replace East-to-West trade routes between New England and Southern Europe as the Yankees' most well-traveled Atlantic highways, and the principal maritime commercial function of the New England cod fisheries in the Atlantic economy became provisioning slave plantations.

Masters of eighteenth-century New England vessels bound to familiar West Indian ports could, therefore, fall asleep on trade voyages and dream of women instead of constantly plotting new courses and tacking different directions to reach novel destinations. The routinization of overseas trade between New England and the West Indies that resulted from the shift in New England's Atlantic cod trade did not come without imperial political costs, however. As subsequent chapters will demonstrate, heads of the British state made political decisions based on this shift.

[100] James L. Howard, *Seth Harding, Mariner: A Naval Picture of the Revolution* (Yale University Press, 1930), 4–5.

5

Atlantic Business Competition and the Political Economy of Cod: Part One

Following the Sugar Act's passage in 1764, a colonial pamphleteer appealed to Parliament "to point out the method of reconciling the present seemingly different and clashing interests of its southern [i.e. West Indian] and northern [i.e. New England] colonies.... That the northern colonies (in which the operation of the said act was chiefly to take place) will be extremely prejudiced in their general navigation, fishery and commerce, is so very obvious."[1]

Edward Payne was more than a little anxious in 1763. He was a fish merchant who lived in Boston. He owned wharfspace, vessels, and a flakeyard in Gloucester, Massachusetts, most likely due to the expense and overcrowding associated with Boston's busy harbor. Payne proudly remarked that his "Own Vessels caught more Fish than the generality of the Vessels in that Town did." He was also a leading member of the "Society for Encouraging Trade and Commerce within the Province of the Massachusetts Bay," a Boston-based merchant consortium devoted to promoting colonial businesses through paid political lobbyists. In this capacity, he had "procured from all the principal Fishing Towns in this Province as exact an account as could be taken of the number of Vessels employed in the Fishery in this Province."[2] This information included how much

[1] "Considerations Upon the Act of Parliament, Whereby A Duty is laid of six Pence Sterling per Gallon on Molasses, and five Shillings per Hundred on Sugar of foreign Growth, imported into any of the British Colonies" (Boston: Printed and Sold by [Benjamin] Edes and [John] Gill, in Queen Street, 1764), 6. Early American Imprints, Series I: Evans #9625.

[2] Sworn testimony of Edward Payne of Boston, Merchant, Boston, December 1763, Ezekiel Price Papers, 1754–1785, MHS. According to the Society's "Rules and regulations," it was meant to "consist of Merchants and others concerned in Commerce, and of any

fish had been caught on an annual basis in Massachusetts; the type of fish that was able to be cured (merchantable or refuse grade); and where the fish was sold. All of this commercial data was then shipped overseas to the colony's lobbyist in London in order to convince Parliament not to pass the Sugar Act, which was then in the process of being debated.

Payne, like the other colonial fish merchants in the society, believed the Sugar Act would threaten the very existence of New England's commercial fishing industry. He explained:

> This Valuable Branch of our Trade & nursery for Sea Men, the Fishery, almost if not wholly depends on our Trade to the Foreign Islands in the West Indies, as we can't cure Fish for the European Market separate from the other sort we send to the West Indies that being only such as is over Salted, Sun Burned, and broken, & thereby rendered unfit for any Market in Europe, and our own islands can't consume [more] of this sort of Fish then [what] is made by the small boats in the Bays, & we have no other Market for what is made by the Bankers, it will be lost if not sent to the Foreign Islands in the West Indies. This loss must infallibly destroy the whole Bank Fishery.[3]

Dried, salted cod represented New England's most significant contribution to the Atlantic economy. Thus, the very threat in 1763 that Parliament would vote the Sugar Act into law was every bit as nerve-racking to Payne and other colonial fish merchants as the actual Act that became law in 1764.

Chapter four demonstrated that the volume and direction of New England's Atlantic cod trade had shifted in the eighteenth century toward a greater concentration on West Indian markets. This chapter will demonstrate that as New Englanders shipped increasing amounts of dried cod to the West Indies, they came to trade more and more outside the boundaries of the British Empire, particularly with French sugar producers. It will also establish that such commerce had political consequences such as the Sugar Act.

Yankees developed particularly strong commercial relations with French buyers in the West Indies during the eighteenth century. New England fish merchants could connect with these buyers through correspondence or through ship captains and supercargoes who traveled to the French

other Persons of Ability and Knowledge in Trade who may be desirous to encourage the same." Rules and regulations of the Committee for Encouraging Trade & Commerce, Ezekiel Price Papers, 1754–1785, MHS. The Society's membership role includes notable Bostonian surnames such as Faneuils, Hutchinsons, and Boylstons.

[3] In the Preamble to a late Act of Parliament, Ezekiel Price Papers, 1754–1785, MHS.

islands. The French government, however, established severe penalties in mercantilist efforts to deter French sugar and sugar products from entering British colonies. France prohibited foreign trade with its West Indies in 1717. Foreign merchants who violated this general prohibition were liable, after a set of laws passed in 1727, to have their vessels and cargoes confiscated. French merchants who engaged in illicit trade even risked being sentenced to labor on Mediterranean slave galleys. Trade with foreign merchants was made legal in 1764, but French planters were strictly forbidden to import dried cod from New England.[4] According to Payne, "the French are so sensible of the disadvantage of selling us their Sugars that they have themselves Prohibited the Trade, under very severe Penalty, & great care is taken in their islands to prevent any white sugars from being brought away least we should supply the European Markets."[5]

As a result of these restrictions, it was easier for Yankees to peddle refuse cod to French buyers in the free port of St. Eustatius. Colonial fish merchants frankly admitted that "most of our Fish & other produce is sold at Eustatius, from whence the French smugglers carry it to their Islands & bring Molasses & Sugars in return, and by means of this Clandestine Trade, we not only vend our own produce, which the English Islands can't take off, but sometimes large quantities ["of English goods" crossed out] Prize Rum at Barbados."[6] St. Eustatius, the "Golden Rock" of the West Indies, was a popular Dutch free port, and those selling New England fish commonly found ready French buyers there throughout the eighteenth century.[7]

[4] Dorothy Burne Goebel, "The 'New England Trade' and the French West Indies, 1763–1774: A Study in Trade Policies," *WMQ*, 3rd Series, Vol. 20, No. 3 (July 1963), 332, 335–336. The focus of this article is on French attempts to maintain a mercantilist economic policy. It does not cover the political economy of cod. For more on French colonial economies, see Paul Butel, *Histoire des Antilles françaises: XVIIe-XXe siècles* (Paris: Perrin, 2002); and Alain-Philippe Blérald, *Histoire économique de la Guadeloupe et de la Martinique du XVIIe siècle à nos jours* (Paris: Karthala, 1986).

[5] In the Preamble to a late Act of Parliament, Ezekiel Price Papers, 1754–1785, MHS.

[6] A loose document numbered 73 at the top and 287 at the bottom, Ezekiel Price Papers, 1754–1785, MHS.

[7] Benjamin W. Labaree, William M. Fowler, Edward W. Sloan, John B. Hattendorf, Jeffrey J. Safford, and Andrew W. German, *America and the Sea: A Maritime History* (Mystic, CT: Mystic Seaport Museum Publications, 1998), 134–135; Ronald Hurst, *The Golden Rock: An Episode of the American War of Independence, 1775–1783* (Annapolis, MD: Naval Institute Press, 1996); Barbara W. Tuchman, *The First Salute: A View of the American Revolution* (New York: Alfred A. Knopf, 1988); and J. Franklin Jameson, "St. Eustatius in the American Revolution," *American Historical Review*, Vol. 8, No. 4 (July 1903), 683–708.

Benjamin Pickman and Samuel Gardner, fish merchants in Salem, Massachusetts, helped compile the 1763 financial data for the Society for the Encouragement of Trade and Commerce.[8] Based on customs records for vessels entering and clearing "Boston & Salem from Jan. 1762 to Jan. 1763," Pickman and Gardner calculated that 266 vessels cleared these Massachusetts ports, bound for the West Indies with cargoes consisting of hogsheads of dried, salted cod, barrels of wet, pickled fish, and lumber. 172 vessels then entered the Massachusetts ports from the West Indies. A total of 2,513 hogsheads of sugar were imported into Boston alone. 1,260 of these hogsheads, or fifty per cent, came from foreign islands. 7,204 hogsheads of molasses were imported. 2,600 of these hogsheads, or thirty-six per cent, entered from foreign islands. A total of 444 hogsheads of West Indian rum entered Boston. Of this total, twenty-seven hogsheads cleared from "Foreign Islands," or six per cent.[9] Only aggregate figures are given for Salem. More molasses and sugar were imported than rum because of New England's domestic rum industry.[10] After being asked to "certify" their data, Pickman and Gardner responded in a second letter: "We think the enclosed List will at once convince any Person... how insufficient our own Islands are to supply us; and we think there cannot be a plainer Proof than that Year will afford, for we had then as good a Right to trade to one Island as another."[11]

The French possessions were particularly attractive among the "Foreign Islands" Pickman and Gardner alluded to in their calculations. Opportunistic New England businessmen saw certain economic advantages to trade with the principal competitors of British West Indian planters. In particular, French planters offered the lowest prices for sugar and its by-products in the Atlantic world. Throughout the eighteenth century, the price differential between British and French West Indian sugar ranged

[8] Benjamin Pickman and Samuel Gardner of Salem, to John Rowe, Esq., and others Committee of the Society for Encouraging Trade & Commerce located in Boston, letter dated Salem, December 24, 1763, Ezekiel Price Papers, 1754–1785, MHS. The data was certainly compiled for political purposes. However, the numbers do not appear to be unduly skewed one way or the other.

[9] The standard hogshead in the late eighteenth century British Empire contained 52.5 gallons. Roger Morriss, "The Supply of Casks and Staves to the Royal Navy, 1770–1815," *The Mariner's Mirror*, Vol. 93, No. 1 (February 2007), 50, footnote 8.

[10] See the discussion in chapter four.

[11] Benjamin Pickman and Samuel Gardner of Salem to Edward Payne and Thomas Gray in Boston, Salem, January 10, 1764, Ezekiel Price Papers, 1754–1785, MHS. By 1762, the British had captured French sugar islands Guadeloupe and Martinique and incorporated them into the formal imperial trading network.

between twenty-five and thirty-three per cent in favor of the French.[12] This comparative price advantage resulted from several factors. France's possessions enjoyed certain natural advantages, including fertile soil and better access to water resources with which to power mills and irrigate fields. The French islands also maintained a more elaborate tax system that financed roads, canals, port facilities, and scientific research pertaining to improving plant variety and processing techniques. As a result, the French islands became increasingly productive throughout the eighteenth century.

By 1770, the combined output of French sugar plantations exceeded that of rival British plantations.[13] In 1775, the plantations on St. Domingue (Haiti) alone surpassed the production of all the British islands combined.[14] Increased production made French West Indian sugar readily available and relatively inexpensive, which provided French planters with a comparative advantage over their British business rivals in the Atlantic marketplace.

French molasses could also be easily had, and for low prices. Molasses was a by-product of the process of converting raw cane into commercial grade sugar.[15] French planters were able to produce a high grade sugar by draining off more of the molasses through the production process.[16] They, therefore, had more molasses to sell than their British counterparts. Moreover, French planters were somewhat desperate to vend this particular by-product, as their choice of markets became greatly restricted over the course of the eighteenth century. At the beginning of the 1700s, the French government prohibited the importation of molasses into France, an enormous market with one of the largest populations in all of Europe. This prohibition was meant to protect the domestic brandy industry from

[12] Andrew Jackson O'Shaughnessy, *An Empire Divided: The American Revolution and the British Caribbean* (University of Pennsylvania Press, 2000), 60–65; Richard S. Dunn, *Sugar and Slaves: The Rise of the Planter Class in the English West Indies, 1624–1713*, 2nd ed. (New York: W.W. Norton & Company, Inc., 1973), 205; and Charles M. Andrews, "Anglo-French Commercial Rivalry, 1700–1750: The Western Phase, II," *American Historical Review*, Vol. 20, No. 4 (July 1915), 763.

[13] Robin Blackburn, *The Making of New World Slavery* (London: Verso Press, 1997), 432, 438; Charles M. Andrews, "Anglo-French Commercial Rivalry, 1700–1750: The Western Phase, I," *American Historical Review*, Vol. 20, No. 3 (April 1915), 550–551.

[14] O'Shaughnessy, *An Empire Divided*, 60–61.

[15] The most comprehensive description of the process involved in early modern sugar production that led to molasses can be found in Stuart B. Schwartz, *Sugar Plantations in the Formation of Brazilian Society: Bahia, 1550–1835* (Cambridge University Press, 1985), 98–131.

[16] Blackburn, *The Making of New World Slavery*, 434.

competition.[17] Before the Seven Years' War, the French Islands could sell molasses to Canadian markets. However, as Payne wrote in 1764: "Molasses is a Commodity the French have no Market to vend it at since the Reduction of Canada. Therefore, they allow us to bring our Fish & lumber to their Islands to purchase it. We want this Article to support our Fishery & carry on our other Trade."[18] The shrinkage of available markets made French planters very willing to sell to New England buyers, and the pressures of supply and demand continually drove down the price of French West Indian molasses.

The French comparative price advantage contributed to cod's extended purchasing power in their ports. Between 1700 and 1775, the price of New England refuse cod ranged between 3.9 to 11.3 shillings per quintal.[19] Between 1720 and 1775, the price of West Indian muscovado sugar, for which there are the most extensive price records, ranged between ten and forty-five shillings per hundredweight in the islands.[20] All things being equal, then, 100 quintals of refuse-grade, dried cod, at a median rate of 7.6 shillings per quintal, would have been worth 760 shillings in the West Indies for much of the eighteenth century. However, because of the price differential that existed between British and French sugar at this time, New England fish merchants could buy more sugar in the French islands. Assuming a median price differential of twenty-nine

[17] O'Shaughnessy, *An Empire Divided*, 62; and Harold Adams Innis, *The Codfisheries: The History of an International Economy* (Yale University Press, 1940), 175–176. According to Andrews, neither molasses nor rum was in great demand in eighteenth-century France: "Molasses was not palatable to the French taste, and the French people would not use it as food, so that the French island planters were compelled to give it to their horses or pigs or to throw it away, while rum was not wanted, because it was too raw a liquor for drinking purposes, and was discouraged because it competed when used with wines and brandies, which ranked high among French staples." Andrews, "Anglo-French Commercial Rivalry, 1700–1750: The Western Phase, I," 556.

[18] "In the Preamble to a late Act of Parliament," 1764, attributed to Edward Payne, Ezekiel Price Papers, 1754–1785, MHS.

[19] Daniel Vickers, "'A Knowen and Staple Commoditie': Codfish Prices in Essex County, Massachusetts, 1640–1775," *Essex Institute, Historical Collections*, Vol. CXXIV (Salem, MA: Newcomb and Gauss, 1988), 198–202, table 1. John J. McCusker has recently reminded us that we must take these cod prices with a grain of salt, as weights and measurements were not standardized around the Atlantic world, and because there were variations in coinage in use throughout the early modern period. See McCusker's remarks in "Roundtable Reviews of Peter E. Pope, *Fish into Wine: The Newfoundland Plantation in the Seventeenth Century* with a Response by Peter E. Pope," in *IJMH*, Vol. 17, No. 1 (June 2005), 251–261.

[20] John J. McCusker and Russell R. Menard, *The Economy of British America, 1607–1789* (University of North Carolina Press, 1985), 157–159, esp. figure 7.1.

per cent and a median sugar rate of 27.5 shillings per hundredweight, 100 quintals of refuse-grade, dried cod could be exchanged for slightly more than twenty-one hundredweights of sugar in the British islands, while the same amount of cod could fetch almost twenty-eight hundredweights in the French islands. In this example, New England merchants could purchase thirty-three per cent more sugar with the French planters than they could otherwise have done with the British planters using the same amount of fish in exchange. Such purchasing power would have then carried over to by-products manufactured from sugar such as molasses.

French West Indian markets for refuse-grade, dried cod were also expanding at a faster pace than rival British markets during the eighteenth century. To be sure, Great Britain gained control over French territory in the West Indies through the course of several wars, and British planters owned more slaves than their French counterparts before the American Revolution (434,000 compared to 393,000 in 1770). However, the slave population on the French islands expanded 4.2 times between 1700 and 1740, and then again by 1.7 times between 1740 and 1770. By comparison, the slave population on British islands only increased 2.2 times and 1.7 times respectively.[21] In other words, between 1700 and 1770, the slave population of the French islands grew 7.3 times (from 54,000 to 393,000), while the same population of the British islands only increased 3.8 times (from 115,000 to 434,000). For New England fish merchants, these slaves represented prospective consumers while their owners were potential buyers. Moreover, in terms of total land mass, the surface area on French islands available for agricultural production and settlement was twice that of the British islands.[22] For New England sellers, these factors meant French markets had the potential to produce even more sugar and import additional slaves above and beyond Great Britain's capacity. Entrepreneurial Yankees able to secure connections and establish networks of trust early on would be almost guaranteed attractive future returns from such burgeoning French markets, which must have appealed to colonial entrepreneurs already operating a risky maritime business.

For their part, wealthy British sugar planters were not content to sit quietly and watch New Englanders sail away with fish and credit to

[21] Stanley L. Engerman, "Economic Growth of Colonial North America," in John J. McCusker and Kenneth Morgan, eds., *The Early Modern Atlantic Economy* (Cambridge University Press, 2000), 236, table 9.1A.

[22] Blackburn, *The Making of New World Slavery*, 432, 438–439.

French competitors. In the first half of the eighteenth century, when French sugar first gained its comparative price advantage, the British planters were upset that New England fish merchants were vending their goods among the French islands and purchasing their chief competitors' sugar products. In 1701, Governor Christopher Codrington of the Leeward Islands recommended Parliament pass a law prohibiting New Englanders from shipping fish to the French West Indies.[23] Henry Worsley, the governor of Barbados, one of the most productive sugar islands, complained to the Board of Trade in 1723 that Yankees were exchanging fish for French molasses to feed their expanding number of distilleries.[24] Governors such as Worsley and Codrington even attempted to seize and condemn colonial vessels found to be trading with French islands.[25] It did not matter that such trade had been perfectly legal as long as it was conducted in British vessels.[26] British sugar interests were powerful, and they demanded that the British state encourage their business. The assembly of Jamaica observed in 1735: "It is very apparent that the [British] sugar colonies have been long declining and very much want the assistance of the legislature [i.e. Parliament] to put them upon an equal footing with their neighbors, the French."[27]

Those concerned in British West Indian sugar production were well positioned to defend their economic interests at the imperial level. British sugar planters were counted among the wealthiest subjects in the empire. Many were absentee landlords who lived in England. They purchased real estate in many different parts of England and built townhouses in cities along with great houses with formal gardens in the countryside. Such land ownership and extreme wealth provided sugar planters with direct access to political power. They won election to the House of Commons. Planters were also three times as likely as anyone from North America to be given a title to nobility from 1701 to 1776, and they frequently intermarried with traditional English peers. This meant that they gained considerable power in the House of Lords. Over the course of the eighteenth century, there were between forty and sixty members of Parliament at any given time

[23] Andrews, "Anglo-French Commercial Rivalry, 1700–1750: The Western Phase, I," 549–550.

[24] Governor Worsley to the Board of Trade, March 26, 1723, *Calendar of State Papers, Colonial Series, America and the West Indies, 1721–23*, No. 486.

[25] Andrews, "Anglo-French Commercial Rivalry, 1700–1750: The Western Phase, II," 764.

[26] Oliver M. Dickerson, *The Navigation Acts and the American Revolution* (University of Pennsylvania Press, 1954), 83.

[27] NA CO 137/21/217–218.

who were either sugar planters themselves or were financially involved in the West Indian plantations. William Beckford, for example, was born in Jamaica, the largest source of sugar in the eighteenth-century British Empire, and he inherited his father's sugar plantation. Beckford grew up and lived in London, where he was twice lord mayor, in 1762–63 and 1769–70. He was a member of the House of Commons between 1747 and 1770, and he was a close personal friend of William Pitt the Elder. Among the seventy-five members of the Board of Trade, a group that advised the crown on colonial affairs, only five had ties to the Western Hemisphere between its genesis in 1696 and the beginning of the American Revolution, but four of these five were commercially connected to the West Indies.[28] Simply put, the sugar industry had powerful advocates at the helm of the British ship of state. Indeed, Lord North, Great Britain's prime minister during the imperial crisis and a man not easily controlled, referred to the West Indian sugar planters as "the only masters he ever had."[29]

Mercantilist minds in Parliament were also predisposed to favor enumerated articles such as tobacco and sugar over commodities like cod that were not listed in any of the Navigation Acts.[30] To be sure, cod could be traded directly to foreign ports because of its long-standing reputation for bringing solid specie into London coffers. Mercantilist theory insisted that there was finite wealth in the world, and the accumulation of hard currency was the surest way to strengthen a nation's economy and gain economic supremacy over commercial rivals. Hence, cod was valued as a trade good for its ability to siphon specie from Iberian nations. However, the profits from the colonial production of dried, salted cod did not always end up in the mother country, as was the case with tobacco and

[28] Stephen J. Hornsby, *British Atlantic, American Frontier: Spaces of Power in Early Modern British America* (University Press of New England, 2005), 46–47, 60; O'Shaughnessy, *An Empire Divided*, 14–17, esp. table 1; and Jacob M. Price, "Who Cared about the Colonies? The Impact of the Thirteen Colonies on British Society and Politics, circa 1714–1775," in Bernard Bailyn and Philip D. Morgan, eds., *Strangers within the Realm: Cultural Margins of the First British Empire* (University of North Carolina Press, 1991), 399–400. For an alternative view of colonial-core relations that stresses the role informal voluntary associations such as churches and coffee houses played in bringing the interests of colonial merchants in line with imperial policy, see Alison Gilbert Olson, *Making the Empire Work: London and American Interest Groups, 1690–1790* (Harvard University Press, 1992). Olson does not investigate the colonial fishing industry.

[29] *Herald: A Gazette for the Country*, August 23, 1797, cited in Alice B. Keith, "Relaxations in the British Restrictions on the American Trade with the British West Indies, 1783–1802," *Journal of Modern History*, Vol. 20, No. 1 (March 1948), 16.

[30] For more on British mercantilist ideas and policies, see Cathy Matson, *Merchants & Empire: Trading in Colonial New York* (Johns Hopkins University Press, 1998).

sugar. Colonists still traded cod with Iberian ports for solid specie, which was then transported to England to purchase and bring home manufactured goods. This trade certainly increased the amount of hard currency in circulation in England and boosted manufacturing in the mother country, yet West Country merchants dominated Southern European markets, and this colonial triangular trade dwindled over the course of the eighteenth century. Commercial connections between New England and the West Indies (British and foreign islands) expanded at this time, and cod was exchanged for sugar products that were consumed and manufactured into rum in New England.[31] Mercantilist MPs, therefore, had an axe to grind with New England's Atlantic cod trade because in their minds it did not directly benefit the mother country to the extent of other commodities.

Wealthy sugar planters further enjoyed the advantage of being able to afford to hire a group of full-time professional agents to lobby Parliament on their behalf. By 1774, the small islands in the West Atlantic maintained ten permanent agents in London whose sole purpose was to promote and defend the interests of the British sugar industry. One of these men, Stephen Fuller, worked as the agent for Jamaica for thirty years between 1764 and 1794. The London-based Society of West India Merchants provided support for these agents. Because most of the sugar produced in the British West Indies was consumed in England during the eighteenth century, London sugar buyers such as Richard Neave and Beeston Long, chairmen of the Society and governors of the Bank of England, were very interested in any effort to lobby Parliament on behalf of the sugar industry.[32]

New England fish merchants did not have the planters' resources or equal access to political power. Yankee merchants by and large did not own real estate in England. They did not control many seats in Parliament. Nor did they have ties to the Bank of England. Of the hundreds of men who were members of the House of Commons between 1754 and 1774, only two individuals, John Tomlinson and Barlow Trecothick received money to represent New England. They were paid to act for New Hampshire. Five merchants who were members of the Commons maintained casual commercial ties to New England.[33] None of these men were willing to sacrifice their political positions for the colonial cod fisheries.

[31] For the decrease in Iberian trade and the increase in West Indian commerce, see chapter four.

[32] O'Shaughnessy, *An Empire Divided*, 15–16, 60–62.

[33] Price, "Who Cared about the Colonies?" 403–405.

Therefore, New England fish merchants were forced by necessity to rely upon a small number of lobbyists.

By 1774, there were just three active North American agents, compared to ten from the West Indies. Only two of these three, William Bollan and Arthur Lee, were full-time advocates for New England interests. Ben Franklin, the third colonial agent, represented Pennsylvania, Georgia, New Jersey, and Massachusetts, and could not be expected to devote much time to the fishing industry.[34]

Fish was also not as lucrative a trade good as sugar, which further restricted colonial fish merchants' ability to lobby Parliament effectively. The total value of British West Indian sugar, rum, and molasses exported between 1768 and 1772 was £3,910,600. By comparison, the British Atlantic fishing interests realized export revenue of £605,000 over the same time period. By itself, New England fish exports were worth £152,155.[35] William Bollan, the agent for Massachusetts for more than twenty-five years leading up to the Revolution, referred to lobbying Parliament as driving "about the wheels of Business," which took "a considerable Expense." Indeed, at one point Massachusetts merchants paid Bollan £2,585 for four years of "attending the negotiation of Business here" in London. Rotating the "wheels of Business" in the eighteenth century, it seems, required considerable sums of money to grease "negotiations" with members of Parliament.[36] Sugar elites had more grease, and more mechanics to lubricate the system, which meant the New England fish lobby was at a decided disadvantage in any attempt to rotate commercial wheels in a favorable direction.

British West Indian planters thus operated from a position of great strength in the imperial contest for Parliamentary affection in the eighteenth century, and whenever their interests opposed those of the New

[34] Jack M. Sosin, *Agents and Merchants: British Colonial Policy and the Origins of the American Revolution, 1763–1775* (University of Nebraska Press, 1965), 176–177. Sosin acknowledges that technically there were five colonial agents, including Edmund Burke (New York) and Charles Garth (South Carolina). However, Sosin maintains that these two agents put their duties as MPs above their roles as advocates. According to Bernard Bailyn, the colonial agent system began during the 1730s. Bailyn, *The Origins of American Politics*, 2nd ed. (New York: Vintage Books, 1968), 92. The number of colonial North American agents in London seems to have remained constant over the course of the eighteenth century. For example, there were five of these agents in 1731. See, *PDBP*, Vol. 4, 103–120.

[35] Hornsby, *British Atlantic, American Frontier*, 26–28, esp. figure 2.1; and McCusker and Menard, *The Economy of British America*, 160, table 7.3.

[36] William Bollan to the Secretary of the Massachusetts General Court, dated Pall-mall, April 19, 1754, Ezekiel Price Papers, 1754–1785, MHS.

England fish merchants, they usually came out on top. To be sure, the British state had not always opposed colonial economic interests. As early as 1643, Parliament permitted ships carrying salt to sail directly from Europe to New England without stopping in England to pay customs duties.[37] The low-priced salt directly benefited the colonial fishing industry. In addition, over the course of the eighteenth century, the British state ratified certain treaties that benefited the New England fishing industry, such as the Treaty of Utrecht in 1713 and the Treaty of Paris in 1763. These settlements severely limited the ability of the French to compete in the Atlantic cod trade, and effectively promoted the operations of colonial British fishermen, both in terms of additional fishing waters and expanded Atlantic markets.[38]

Nevertheless, the West Indies enjoyed a longer and stronger tradition of support from the British state. In the Treaty of Breda in 1667, Charles II ceded Nova Scotia to the French in exchange for Antigua, Montserrat, and the rest of St. Christopher's island. Colonists in Massachusetts protested, citing the threat nearby French competitors posed to their fishing operations. A premium, however, was placed on sugar production because of the revenue it generated.[39] In addition, a 1748 act exempted the British West Indies from naval impressment. The same act did not mention the New England colonies despite their appeals for formal renewal of the "Sixth of Anne," which prohibited the naval impressment of colonial mariners.[40] Furthermore, between the Treaty of Breda and the American

[37] Brian Fagan, *Fish on Friday: Feasting, Fasting and the Discovery of the New World* (New York: Basic Books, 2006), 284.

[38] See the discussion in chapter four.

[39] Bernard Bailyn, *The New England Merchants in the Seventeenth Century*, 3rd ed. (Harvard University Press, 1995), 128.

[40] Richard B. Morris, *Government and Labor in Early America* (New York: Harper & Row, Publishers, 1946), 276. 6 Anne c. 37, Great Britain, *Statutes at Large from the Tenth Year of King William the Third to the End of the Reign of Queen Anne*... Vol. 4 (London, 1769), 336. The Sixth of Anne formally stated: "That no mariner, or other person, who shall serve on board, or be retained to serve on board, any privateer, or trading ship or vessel that shall be employed in any part of America, nor any mariner, or other person, being on shore in any part thereof, shall be liable to be impressed or taken away by any officer or officers of or belonging to any of her Majesty's ships of war, empowered by the Lord High Admiral, or any other person whatsoever, unless such mariner shall have before deserted from such ship of war belonging to her Majesty, at any time after the 14th day of February 1707, upon pain that any officer or officers so impressing or taking away, or causing to be impressed or taken away, any mariner or other person, contrary to the tenor and true meaning of this act, shall forfeit to the master, or owner or owners of any such ship or vessel, £20 for every man he or they

Revolution, the British state established a series of prohibitive duties on foreign sugars, guaranteeing planters a permanent monopoly in the home market.

The British state passed the Molasses Act in 1733 despite opposition from the New England fishing interests.[41] The "act for the better securing and encouraging the trade of his Majesty's sugar colonies in America" set prohibitive duties on foreign sugar ("Five Shillings per Hundred weight"), molasses ("Six pence per Gallon"), and rum ("Nine pence a Gallon"), which, if enforced and obeyed, would have made New England trade with the French West Indies less profitable.[42] In the preamble to the Act, Parliament justified its passage in the following manner: "Whereas the welfare & prosperity of your Majesty's sugar colonies in America are of the greatest consequence and importance to the trade, navigation and strength of this kingdom; and whereas the planters of the said sugar colonies, have of late years, fallen under such great discouragements that they are unable to improve or carry on the Sugar Trade, upon an equal footing with the Foreign Sugar colonies, without some advantage and relief given to them from Great Britain."[43] The Act could have given British planters a second monopoly, in addition to the one over markets in England, this time on the colonial North American markets. The

shall so impress or take, to be recovered with full costs of suit, in any court within any part of her Majesty's dominions."

[41] *PDBP*, Vol. 4, 162–163, esp. footnotes 82–86. Yankee merchants and their agent protested that the markets among the British islands were too small to consume all of the fish sent south, and that these same markets could not provide enough supplies to fill holds on the return trip north. For more arguments against the Act, see *The Case of the Provinces of the Massachusetts Bay*... (London: 1731); and *The Case of the British Northern Colonies* (London: 1731). For arguments in favor of the Act, see *Observations on the Case of the British Northern Colonies* (London: 1731); and *The Importance of the British Sugar Plantations in America to this Kingdom*... (London: 1731). For more on the Molasses Act, see Richard B. Sheridan, "The Molasses Act and the Market Strategy of the British Sugar Planters," *Journal of Economic History*, Vol. 17, No. 1 (March 1957), 62–83. The West Indian sugar interests had a similar bill rejected in 1731. Charles McLean Andrews argued that because British planters and their friends failed to get their protective legislation passed on the first attempt, Parliament did not favor their interests. He also cited the return of Guadeloupe to France in 1763 as further evidence of Parliament's disfavor. Andrews, "Anglo-French Commercial Rivalry, 1700–1750: The Western Phase, II," 772–773. Comparatively, Parliament greatly favored the sugar interests over the New England fishing interests.

[42] For a copy of the act, see *New England Weekly Journal*, December 31, 1733.

[43] The 1733 Act was republished in *New-Hampshire Gazette, and Historical Chronicle*, January 29, 1768.

Molasses Act was not enforced, however, and colonial fish merchants found various ways around the letter of the law.[44] Yet, the law effectively turned legitimate colonial entrepreneurs with genteel pretensions into smugglers, and continual pressure from the sugar interests resulted in the renewal of the Act every five years until 1764.[45]

New England fish merchants and sugar planters continued to lobby against one another after 1733. Not satisfied with the Molasses Act, the sugar lobby demanded stricter enforcement and more prohibitive duties almost from the moment the bill became law. Colonial fish merchants, by contrast, did not want any restriction on their ability to sell fish. In 1754, Bollan wrote to the Massachusetts General Court to inform its merchant members that the "West Indians," whom he referred to as "the adversaries," were "indefatigable, sparing no pains or Expense of any kind, but use their utmost Endeavor in every shape to carry their point." The representative of the colonial fishing industry was optimistic at this time and expected a resolution to the Act that would at last result in New England's favor: "[W]e have thus far been able to defeat all their late Attempts, and as we have from time to time gained Ground of them, there is in my humble opinion, the greatest reason to go on, and prepare for the approaching and final Trial and Determination of this Matter."[46] However, renewed conflict with the French and the ensuing Seven Years' War delayed the "final Trial" Bollan anxiously anticipated. The war also altered the outcome Bollan expected in ways he could not have foreseen.

The Seven Years' War was one of the first truly global conflicts, and it stretched the limits of Great Britain's military and fiscal abilities.[47] The war embroiled European powers in combat in the Atlantic, Indian, and Pacific oceans. The British government borrowed enormous sums of money to equip and provision troops, to purchase and maintain

[44] For more on the extent to which the Act was evaded, see Thomas C. Barrow, *Trade and Empire: The British Customs Service in Colonial America, 1660–1775* (Harvard University Press, 1967), 134–143; and Dickerson, *The Navigation Acts and the American Revolution*, 66–69, 82–87. Dickerson has gone too far in arguing that, because of widespread colonial evasion, neither the Molasses Act nor the Sugar Act had much of an impact in terms of motivating colonists to resist British authority.

[45] One sympathetic MP later remarked in the midst of Parliamentary debates, "[O]ur laws have created smuggling even by force. Smugglers of molasses instead of being infamous are called patriots in North America." *PDBP*, Vol. 1, 490.

[46] William Bollan to the Secretary of the Massachusetts General Court, Pall-mall, April 19, 1754, Ezekiel Price Papers, 1754–1785, MHS.

[47] According to Jonathan Israel, the Second Anglo-Dutch War (1664–1667) was a global war. Jonathan I. Israel, *The Dutch Republic: Its Rise, Greatness, and Fall, 1477–1806*, 4th ed. (Oxford University Press, 1998), 766–774.

weaponry, and to build and outfit warships. Britannia emerged victorious over seas as well as lands, but it cost the nation £160 million just to fight the war – twice the gross national product. Britain was just wealthy enough to pay its creditors £14 million at the end of the war, leaving a huge debt for the future. Maintaining garrisons in North America alone cost £200,000 per year during the 1760s.[48] The immense war debt, coupled with additional military expenditures carried over from the war, left the heads of the British state desperately looking for ways to raise additional revenue.

As a result of this war debt, Parliament debated in 1763 whether or not to make the Molasses Act permanent, and what, if any, alterations needed to be made to the Act to give it teeth.[49] It was known in the colonies that Parliament was debating the issue, and the Society for the Encouragement of Trade and Commerce busily prepared for the 1763 debate. Members of the Society corresponded with "Merchants of Salem, Marblehead, and Plymouth," convened "diverse meetings of committees from the several Towns," and designed "Reasons against the Renewal of the Sugar Act," which were "presented to the General Court . . . were adopted by them, and transmitted to their agent" in London. In this manner, the Society played an important role in facilitating the colony's lobby efforts. At great expense, the Society "also sent copies [of their "Reasons"], fairly wrote, to the Merchants of six of the Neighboring Governments, and afterwards 250 printed copies to the Principal Merchants in London."[50]

The fish merchants also made public their grievances in a Boston newspaper. In a series of letters, a colonial dissenter writing under the pseudonym "Baron de Montesquieu," a reference to the French Enlightenment figure, laid out all of the arguments that Massachusetts's colonial agent had been putting before Parliament. In addition, "Montesquieu" reminded colonists that the "seasons for curing fish are some years so unfavorable, that the greatest part of what is caught is only fit for the West India market. . . . The chief part of this fish we depend on the foreign sugar

[48] Fred Anderson, *Crucible of War: The Seven Years' War and the Face of Empire in British North America, 1754–1766* (New York: Vintage Books, 2000), 562; Nancy F. Koehn, *The Power of Commerce: Economy and Governance in the First British Empire* (Cornell University Press, 1994), 5, 12; and John Shy, *Toward Lexington: The Role of the British Army in the Coming of the American Revolution* (Princeton University Press, 1965), 241.

[49] For the debates on the "Sugar Bill" in Parliament in 1763, see *PDBP*, Vol. 1, 450–451.

[50] Thomas Gray and Edward Payne, "To the Members of the Society for Encouraging Trade & Commerce," Boston, January, 1766, Ezekiel Price Papers, 1754–1785, MHS.

colonies to take off our hands. The only pay they have to give, therefore, is sugar and molasses; an article of great use to fishermen." The dissenter who dared not publish his real name believed, "A reformation [in trade regulations] is become necessary. But should such restraints take place as to discourage the trade, and destroy the fishery of the northern colonies; Great Britain, instead of benefiting itself by its conquests [i.e. the Seven Years' War], has only added a load which it must at length sink under." In short, "Montesquieu" argued that a permanent tax on foreign sugars would disrupt the trade in refuse grades of dried cod that were shipped to the West Indies. If this trade were disrupted, he reasoned, then the conquest of Canada and Newfoundland had been for naught. "Montesquieu" attributed the existence of taxes on foreign sugars, and the plans to make them permanent and more rigidly enforced, to the powerful British sugar interests, or "To the aggrandizement of the planters of our sugar colonies; who, notwithstanding they have become so opulent, that upwards of forty of them have seats in the House of Commons, and numbers have formed alliances with some of the noblest families in the kingdom, do yet complain for want of further encouragement. *Oh fatal Policy! Error irretrievable!*"[51]

In many ways, debates that occurred in Parliament in 1763 over the Sugar Act represented a continuation of the commercial rivalry that existed between the New England fish merchants and the sugar interests since the French lowered their West Indian prices in a deliberate attempt to seize Atlantic market share. Once again, those representing the fishing industry argued that British West Indian markets could not consume all of New England's fish, nor could British planters provide enough sugar products in exchange. Yankee agents reminded Parliament that the profits from their trade with the French ended up being used in exchange for manufactured goods from Great Britain. Thus, restricting New Englanders' profits would ultimately hurt domestic industries in England. The agents also tried to convince Parliamentarians that colonial fishing vessels and maritime workers contributed to the "nursery" for naval seamen, and, therefore, the encouragement of their fisheries was closely linked to matters of military preparedness and state power.[52] These were the same arguments agents representing the interests of the colonial fisheries had

[51] *Boston Evening Post*, November 21, 1763; *Boston Evening Post*, November 28, 1763.

[52] "In the Preamble to a late Act of Parliament," 1764, attributed to Edward Payne, Ezekiel Price Papers, 1754–1785, MHS. Also see, *Reasons Against the Renewal of the Sugar Act* (Boston, 1764), Early American Imprints, Series I: Evans #9812. For more on the English belief that commercial fisheries represented nurseries of seamen, see the discussion in chapters six and eleven.

trotted out for decades. These same arguments would be wielded again in the 1770s, and to the same effect.

Despite fish merchants' best efforts to publicize their arguments at home and abroad, Parliament ultimately found in favor of the sugar interests, just as it had done in 1733 and on other occasions. The Sugar Act, also known as the Revenue Act, became the law of the empire in 1764. It was part of an imperial package of revenue-raising measures instigated by Lord Grenville to pay off Seven Years' War debts. It levied a tax of "[o]ne pound, two shillings, over and above all other duties imposed by any former act of parliament" on foreign white sugars imported into the colonies. The importation of foreign rum was forbidden, giving British West Indian planters a monopoly on the rum trade with North America. To be sure, Parliament reduced the incentive to smuggle by immediately lowering the duty on foreign molasses to "three pence." Such a move can be construed as a concession to the New England fishing interests, yet New Englanders were not satisfied as they desired no tax whatsoever. Moreover, the duties on sugar and molasses were made "perpetual" starting in September 1764.[53]

The Sugar Act did not simply restrict North American trade to the West Indies. To prevent direct commerce with French possessions anywhere in the Atlantic, and thereby restrict North American access to inexpensive foreign sugar products, the Sugar Act expressly forbade colonial trade with French territories in the far northern Atlantic. As part of the 1763 peace treaty ending the Seven Years' War, France retained only a few islands off the south coast of Newfoundland (St. Pierre and Miquelon) and fishing rights to the western shoreline of the land of cod.[54] These islands, however, were important to New England commerce, and they had been for some time. As early as 1684, New Englanders were supplying French fishermen and traders at St. Pierre with salted cod and rum in exchange for West Indian goods, European manufactured products, and lines of credit.[55] Thus, it was no small matter that the 1764 Act explicitly prohibited direct trade with "the subjects of the crown of France in the islands of Saint Pierre and Miquelon."[56]

[53] *PDBP*, Vol. 1, 492–494; *Anno regni Georgi III*... (London, 1764), Early American Imprints, Series I: Evans #41449; and *Boston Evening Post*, June 25, 1764.

[54] Clive Parry, ed., *The Consolidated Treaty Series*, Vol. 42 (New York: Oceana Publications, Inc., 1969), 320–345.

[55] Innis, *The Codfisheries*, 122.

[56] *Anno regni Georgi III*... (London, 1764), Early American Imprints, Series I: Evans #41449, 17. Lord Grenville, the head of the Treasury Department and the driving force behind efforts to reform colonial revenue collection, argued in 1764 that such a

To give the Sugar Act teeth, Parliament restructured custom collection and gave new orders to the Royal Navy. Absentee customs collectors, who had been collecting little but their stipends, were ordered to return to their posts. Officials who did not comply were removed. Informants who aided custom collection received greater rewards. Customs officials who cooperated with smugglers faced harsher punishment. The Act further created additional customs offices and added manpower to these offices. It also required traders at the port of departure and at the port of entry to carry new certificates signed by justices of the peace designating the origin and quantity of all sugar products on board. Shippers were required to post additional bonds as assurance that their goods would be landed only in British ports. Ships leaving colonial ports also had to have a "cocket," or cargo list. This list could only be obtained from the customs house, and it catalogued which goods were subject to import and export duties and which duties had been paid. Even sailors' private ventures were required by law to be included on the cocket and were subject to duties. The Act directly ordered royal governors to be more vigilant in the collection of His Majesty's customs. Royal Navy patrols were more active and more numerous. Hence the HMS *Rose's* patrol off Marblehead and the *Pitt Packet* affair. Naval officers such as Captain Benjamin Caldwell viewed themselves as enforcers of martial law on the high seas. And any vessel caught within two leagues of a British port without a cocket was subject to seizure by the navy. These seizures were primarily prosecuted in a special Vice-Admiralty court established in Halifax, Nova Scotia, for the express purpose of trying violators of the Sugar Act.[57]

The navy placed special emphasis on enforcing the Sugar Act in the home waters of the offending New Englanders, and most of its warships were positioned along the Northwest Atlantic littoral between Massachusetts and Newfoundland. Neil R. Stout, the foremost authority on the naval enforcement of British colonial policy, maintains that the concentration of military force was primarily due to two factors: Halifax

prohibition was necessary "to take care that all illicit commerce be stopped with St. Pierre and Miquelon." *PDBP*, Vol. 1, 492.

[57] Neil R. Stout, *The Royal Navy in America, 1760–1775: A Study of Enforcement of British Colonial Policy in the Era of the American Revolution* (Annapolis, MD: The United States Naval Institute, 1973), 25–38, 55, 56–90, 166; Leonard Woods Labaree, ed., *Royal Instructions to British Governors, 1670–1776*, 2 Vols. (New York: Octagon Books, Inc., 1967), Vol. 2, 766–767, 780; and Carl Ubbelohde, *The Vice-Admiralty Courts and the American Revolution* (University of North Carolina Press, 1960), 49–54, 101–103.

represented "the principal naval base" and "the only royal dockyard" in North America, and the British wanted to "overawe the Canadians and impress France."[58] Certainly, these factors cannot be ignored. New England merchants, however, were the principal violators of the Sugar Act, and cod was the primary commodity used in exchange for French contraband. Thus, the British navy likely positioned its forces to cut off illicit trade at its source as well as to intimidate the French.

Indeed, Commodore Hugh Palliser, the naval officer in charge of a portion of the British North Atlantic fleet and the royal governor of Newfoundland between 1764 and 1769, ordered his squadron to seize "all New England Vessels" found prosecuting such "Illicit Trade."[59] In 1765, Palliser reported to the Board of Trade: "Great Numbers of New England Vessels are also become Fishers for the French at those Islands [St. Pierre and Miquelon], and Engaged in Trade with them for French Goods and Manufactories; however, I believe they will not pursue it so Eagerly as they had begun, for we have taken & Confiscated Five of them."[60] According to contemporary estimates, the cost for purchasing a single deep-sea fishing schooner and outfitting a crew for one year was £1,135.[61] In 1755, £1,135 sterling was equivalent to $131,528.34 in year 2000 U.S. dollars.[62] Thus, Palliser's capture of five of these vessels in 1765 corresponds to a financial loss of over $657,000 in 2000 U.S. dollars. That same year, Boston fish merchant John Rowe recorded in his diary: "Mr. [Elbridge] Gerry [a prominent Marblehead fish merchant] came to Town & brought an account of the [HMS] *Niger*, Man of War, taking three Schooners out of the Harbor of St. Pierre's, one belonging to his Father & two of Epps Sargent [a Gloucester merchant]."[63] One year later, a colonial newspaper reported: "They write from Newfoundland, that several New England sloops have been seized by his Majesty's ships of war at Cape

58 Stout, *The Royal Navy in America, 1760–1775*, 57.

59 Palliser's Order "to Mr. Nicholas Gill, Naval Officer at St. Johns, authorizing him to seize Illicit Traders," St. Johns, November 4, 1764, Colonial Secretary Letterbooks, 1752–1765, MUMHA, 276–277.

60 Palliser to the Board of Trade & Plantations in London, St. John's, Newfoundland, October 30, 1765, NA CO 194/16/170–174.

61 "Calculations respecting outfits of a Fishing vessel," December 1763, Ezekiel Price Papers, 1754–1785, MHS.

62 John J. McCusker, *How Much Is That in Real Money? A Historical Commodity Price Index for Use as a Deflator of Money Values in the Economy of the United States*, 2nd ed. (Worcester, MA: American Antiquarian Society, 2001), 35–36.

63 Anne Rowe Cunningham, ed., *Letters and diary of John Rowe: Boston merchant, 1759–1762, 1764–1779* (Boston: W.B. Clarke Co., 1903), 87.

Breton, for smuggling with the French, and selling them large quantities of ready cured fish.... by means of which several fishermen from [La] Rochelle and Bordeaux got loaded in a few days, and had actually sailed home with cargoes of fish ready cured for the Mediterranean markets."[64] Thus, the Sugar Act, which was meant to encourage the British West Indian islands, came at the expense of New England's maritime commerce in general, and its fishing industry in particular.

New England colonists considered the Sugar Act as just one, and the latest, piece of evidence among many that the British state-favored West Indian sugar interests over colonial fishing interests. They reacted to the legislation with bitterness and frustration. One colonial critic published his grievances under the *nom de plume* "Americanus." This writer denounced the Act, as its prohibitive duties on foreign sugar and molasses and its ban on foreign rum effectively denied merchants in "the New England fishery" the "disposal of [refuse-grade cod] to the French and Dutch." Because of the loss of these foreign markets and the denial of the principle of free trade, "difficulties fell upon the New England fishery, [and] in the chief fishing town (Marblehead) the vessels were reduced one half in their number." "All the merchants" in New England, he wrote, believed "that the continuance of the duty on molasses, with other duties imposed by the said act [the Sugar Act], will inevitably destroy" the "fishing trade." Americanus further estimated that New Englanders only imported "one eighth part" of their molasses from the British West Indies because of high prices. Thus, another result of the Act was to limit the amount of West India goods "had by the fishermen."[65]

Distillery owners in colonial New England would have stood in solidarity with the fish merchants in opposition to the Sugar Act. Clearly, the tax on imported molasses increased the cost of doing business for distillery owners. Less obvious, perhaps, is the fact that the Act's threat to the fishing industry would have disturbed these owners as well. This resulted from important final demand linkages that connected the fisheries' fate to the distilleries' destiny.[66] West Indian molasses was frequently purchased with refuse grades of North Atlantic cod. Fish merchants conducted these

[64] *Georgia Gazette*, September 3, 1766.

[65] *Boston Evening Post*, January 20, 1766.

[66] Final demand linkages refer to the use of the income generated through the sale of a staple in another industry at a later date. See McCusker and Menard, *The Economy of British America*, 24–25; and Melville H. Watkins, "A Staple Theory of Economic Growth," *Canadian Journal of Economics and Political Science*, Vol. 29, No. 2 (May 1963), 145.

exchanges and sold the molasses to the distillery owners. Additionally, these owners purchased refuse-grade, dried, salted cod from fish merchants. In this case, the owners transported the fish themselves, saving on freight rates. For example, Salem distillery owner and merchant Richard Derby purchased refuse fish in Massachusetts for sale in the West Indies. John Bartoll of Marblehead sold Derby refuse cod between 1763 and 1764. Thomas Stevens of Marblehead also sold Derby refuse cod in 1756 and 1757. Derby then shipped the fish south on vessels such as the schooner *Kate*, which sailed to St. Eustatius in the winter of 1757.[67] It should not be overlooked that threats to the fishing industry were ominous portents to colonial entrepreneurs with backward linkages to the fisheries, such as shipbuilders, carpenters, painters, shipriggers, sailmakers, *and* entrepreneurs like Derby who maintained final demand linkages to the fisheries.

Colonists lashed out at British West Indian planters, whom they immediately blamed for the Sugar Act. In a "Statement of the Trade and Fisheries of Massachusetts," written in 1764, Boston fish merchants stated their belief that the planters had, "with no other view than to enrich themselves," forced "the northern Colonies to take their whole supply from them; and they still endeavor the continuance of it under a pretence, that they can supply Great Britain and all her Colonies with West India goods, which is perfectly chimerical."[68] A Providence, Rhode Island newspaper mourned the fact that "the Continent must henceforth move in the wake of the sugar planters of the British Empire."[69]

Whiggish Parliamentarians in London agreed with colonists that the sugar interests had won the day. Edmund Burke, one of the most eloquent opposition members of Parliament in the 1760s, stated his belief that the Sugar Act was a deliberate attempt by corrupt heads of state to generate revenue; to provide planters with a monopoly on North American trade; and to enforce this monopoly with the Royal Navy. The Act, in Burke's words, represented "a revenue not substantiated in place of, but superadded to a monopoly, which monopoly was enforced at the same time with additional strictness, and the execution put into military hands."[70]

[67] Richard Derby Ledger, 1757–1776, JDPL. Derby owned fishing vessels and employed fishermen in addition to owning a distillery. He can, therefore, be seen as both a fish merchant and a rum producer.

[68] Cited in O'Shaughnessy, *An Empire Divided*, 66.

[69] *Providence Gazette and Country Journal*, October 27, 1764.

[70] Edmund Burke, "Speech on American Taxation," April 19, 1774, in *Edmund Burke on Government Politics and Society*, B.W. Hill, ed. (London: Harvester Press, 1975), 125.

Some frustrated colonists continued to work within the system to promote change. In 1765, Massachusetts petitioned Parliament "for the repeal of the Sugar Act." This petition maintained "that the importation of foreign molasses into this province in particular, is of the greatest importance." The colonists believed the tax on sugar and molasses constituted a "prohibition" on the trade. They noted that the importance of their ability to import West Indian sugar products "does not arise merely nor principally from the necessity of foreign molasses, in order to its being consumed or distilled within this province." Rather, they maintained "that if the trade for many years carried on for foreign molasses can no longer be continued, a vent cannot be found for more than one half the fish of inferior quality [i.e. refuse-grade, dried cod], which is caught and cured by the inhabitants of this province; the French permitting no fish to be carried by foreigners to any of their islands, unless it be bartered or exchanged for molasses." The colonists further argued "that if there be no sale of fish of inferior quality, it will be impossible to continue the fishery." The combination cure adopted throughout New England in the eighteenth century made it impossible to catch and cure cod without producing a certain percentage of fish that was not saleable in European markets. If New Englanders could only sell to European markets, then they would have to raise their prices to make up for the loss of the West Indian markets. As a result, "the fish usually sent to Europe will then cost so dear, that the French will be able to undersell the English at all the European markets, and by this means one of the most valuable returns to Great Britain will be utterly lost, & that great nursery of seamen destroyed."[71]

There is some evidence to suggest that the reaction of the fishing interests changed Parliament's sentiment regarding the Sugar Act. Eventually, the tax on molasses was lowered. In 1766, the duty was reduced to a single penny per gallon. The representatives of the colonial fishing industry must get some credit for the reduction. However, as Oliver M. Dickerson points out, the change was primarily the result of pressure from business interests associated with the Atlantic slave trade. These interests argued in 1766 that the tax was preventing provisions, including fish, from reaching the West Indian plantations and adversely impacting the market for slaves in the region.[72] Thus, it was this powerful lobby, not the New England fishing lobby, that was the main force in reducing the tax on molasses.

[71] *Boston Evening Post*, March 11, 1765.

[72] Dickerson, *The Navigation Acts and the American Revolution*, 85–87.

Additionally, despite the fact that the duty had been lowered to a level at which trade could still continue with French West Indian markets, the tax remained and grated on Yankee sensibilities. New England had grown used to a policy of "salutary neglect" that gave them a degree of free trade. Indeed, this is precisely why Adam Smith supported the colonists in their grievances with the Empire.[73] Some in the fishing industry thought raising of revenue through taxes without colonists' consent was evidence of a corrupt state attempting to turn its people into subjects without liberties or slaves. Benjamin Bangs, a Massachusetts fisherman who had become a wealthy merchant and slave owner through the West Indian trade, recorded his anger and frustration in his personal diary. Following news that the Sugar Act had passed both houses of Parliament, Bangs wrote on October 1, 1764, "Now we may date our slavery. Boston is all in uproar about it. But can't help it." A few months later, on December 15, Bangs noted that colonists were beginning to feel more and more like un-free men, constantly policed and watched so as to keep their trade fettered. In his words, there were "men [of] war in every port guarding us more in peace than in war."[74] For a slave owner with commercial ties to the West Indies such as Bangs, Parliament seemed to be turning the world upside down by forcing a prosperous merchant into monitored bondage.

Even those colonists predisposed to look on the British state with favor did not appreciate the Sugar Act's curtailment of colonial economic freedoms. John Powell had supplied provisions to British naval vessels patrolling the New England coast for smugglers in the 1760s. His business ties eventually led him to distance himself from the revolutionary movement, yet even this Boston merchant wrote in a letter to a friend that news of the Act was "very Alarming To the Colonies." In Hobessian rhetoric conjuring images of Leviathans, Powell wrote that "the [Sugar] Act will prove an unformed monster;" he continued, "That will Devour all before It & its makers In The End. The principals of a Commercial Nation [i.e. free trade] are subverted." He further considered the Act "An Unconstitutional Force Laid on the American subjects That must finally End in Riot & Discord & be more severely felt In Great Britain Then Even The present Generation can suffer." Like Bangs, Powell believed Parliament had sent the Royal Navy in greater numbers to their coasts, "Doubling the Cutters [patrol vessels]," and had added "an Army of Excisemen" in an attempt to restrict colonial commercial freedoms and

[73] Adam Smith, *An Inquiry into the Nature and Causes of The Wealth of Nations*, Max Lerner, ed. (New York: The Modern Library, 1937; orig, pub. in 1776), 616–617.

[74] Benjamin Bangs Diary, 1742–1765, MHS.

police trade. Unlike Bangs, who thought the Act would convert colonists into slaves, Powell believed, it "will oblige us all To hold the plough, the scythe, and Reaping Hook."[75] In short, the Bostonian who supplied British naval vessels saw the act as an attempt to cut off colonists' access to the sea and maritime ways of life.

John Rowe was born in England in 1715 and emigrated to Boston by mid-century. Like Bangs, he became a very successful fish merchant and slave owner who exported fish to the West Indies. Unlike Bangs, Rowe became a Loyalist during the Revolution, yet on September 29, 1764, the day the provisions of the Sugar Act were to take effect in the colonies, Rowe recorded in his diary, "The Black Act takes place this day."[76]

While Loyalists like Rowe and Powell remained content operating within the British imperial political system, the Sugar Act went a long way toward convincing other colonists who sided more with Bangs that the British imperial political system itself was corrupt and attempting to strip colonists of their liberties and freedoms. At the most basic economic level, the act restricted colonial fish merchants' access to foreign markets through a prohibitive tax. This meant that most of the dried fish processed in New England could not be sold. It also meant that Yankee trade vessels would have to return from the West Indies with only partially filled holds, as the British islands could not produce enough sugar products to meet the needs of England and the colonies. Thus, in the short run, the Act meant reduced profits and increased labor costs for fish merchants. In the long run, such lower profits would have meant greater unemployment in the fishing industry.

The Act further raised political issues of taxation without representation that predated colonial concerns associated with the Stamp Act.[77]

[75] "Letter dated Boston, April 9, 1764," John Powell Letters, 1764–1773, MHS. Fish merchants held a town meeting in Marblehead, Massachusetts, in 1773 shortly after Parliament renewed the tax on tea that had been part of the 1767 Townshend Duties. The language in the meeting's resolves with respect to the tax on tea mirrors Powell's earlier sentiments: "That Americans have a right to be free as any inhabitants of the earth, and to enjoy at all times an uninterrupted possession of their property.... daring attempts upon the liberties of America have justly won the contempt and severest marks of resentment from every American. The brave citizens of Boston are commended for their noble firmness in support of American Liberty, and the people of this town declare that we are ready with our lives and interests [i.e. property] to assist them in opposing these and all other measures tending to enslave our country." Marblehead Town Records, Office of the Town Clerk, Abbot Hall, Marblehead, Massachusetts, December 7, 1773.

[76] Cunningham, ed., *Letters and diary of John Rowe*, 64–65.

[77] For more on colonial reactions to the Stamp Act, see Edmund S. Morgan and Helen M. Morgan, *The Stamp Act Crisis: Prologue to Revolution*, 3rd ed. (University of North Carolina Press, 1995).

In 1764, the elected members of the lower house of Massachusetts's legislature concluded with regard to the Sugar Act that "the Power of Taxing" should be "exercised with great Moderation" when applied to "the Dominions who are not represented in Parliament." They continued, "For this last is the grand Barrier of British Liberty; which if once broken down, all is lost. . . . a People may be free and tolerably happy without a particular Branch of Trade, but without the Privilege of assessing their own Taxes, they can be neither."[78]

The Sugar Act also brought into stark relief the political corruption that seemed to come from the influence and money of the sugar interests. Some believed that traditional efforts to work within the system and lobby Parliament through colonial agents had failed to bring about real, lasting change because of this corruption. Members of the lower house of Massachusetts's legislature collectively queried: "If a West-Indian, or any other bye Influence is to govern and supersede our most essential Rights as British Subjects, what will it avail us to make Remonstrances, or the most demonstrable Representations of our Rights and Privileges?"[79] A similar sense of frustration can be felt in a petition dated November 25, 1772, in which fish merchants in Marblehead, Massachusetts, called for a special town meeting to voice their concerns at the current state of imperial affairs. The petition justified the meeting by explaining that "the complaints of so considerable a part of the Empire as is made up of North America are unattended to; their petitions, remonstrances, and resolves neglected and despised; acts of Parliament contrary to the apparent rights of mankind, and British subjects, *carried in execution and crammed down their throats*."[80] Such frustration spilled over into extra-legal forms of protest.

Chief among the early colonial efforts to change the system by operating outside its legal boundaries was non-importation. Colonial merchants in the northern colonies established non-importation agreements in 1765, immediately following the Sugar and Stamp Acts, and again in 1768, following the Townshend Act. The Boston-based Society for the Encouragement of Trade and Commerce voted on a non-importation agreement in 1768. Without legal sanction, this commercial consortium, which included fish merchants, agreed to boycott British manufactured goods for one year. The consortium exempted only those items,

[78] *JHRM*, Vol. 41, 77.

[79] *JHRM*, Vol. 41, 72.

[80] Marblehead Town Records, November 25, 1772, Abbott Hall, Office of the Town Clerk, Marblehead, Massachusetts. Emphasis in the original.

chiefly salt, deemed necessary for the prosecution of the fishing industry. The Society succeeded in getting other commercial ports to join in the boycott. The agreement was extended for another year, and in 1769 they formed a committee of inspection to enforce the trade embargo in Boston.[81] Such non-importation agreements enabled colonial merchants to sell surplus manufactured goods that had been stockpiling in warehouses, and they sent a clear political message to Parliament that commercial duties would not be tolerated.[82] In one way or another, colonists were beginning to assume the authority of regulating their own trade, thereby asserting their willingness to defend their economic freedoms.

Colonists also engaged in a number of illegal mob activities aimed specifically at His Majesty's customs collection during the 1760s. Those customs agents who conscientiously attempted to enforce revenue acts such as the Sugar Act and Townshend duties were especially targeted in colonial fishing ports. Thomas Row, the tidewaiter for Salem and Marblehead responsible for checking vessels' manifests as they entered and cleared off, was brutalized for informing his superiors that the schooner *Neptune* was bringing foreign molasses into Salem in 1768. Row was "surrounded by a large Mob with a flag flying." Once captured, he wrote that one of the mob "waxed the hair of my head with Balls of Shoemaker's Wax, and some amongst them (I cannot say which) cut part of it off from behind." Row was then taken "to the Commons, striped of my Coat, Waistcoat, handkerchief & hat." According to an eyewitness, "They had tarred his Clothes all over, and then covered him with feathers, & pinioned him into the Cart, and were carrying him through the Town as a Spectacle of Infamy; Stopping at all the Public Offices." Row testified that the crowd "fixed two boughs with spun yarn on each side of my head." They also "put a cross made with sticks into my hand, struck me several violent blows on my head, and swore they would murder me if I did not carry it upright." When the crowd finished with Row, they took him "to the extreme part of the town," let him go, and told "him he must

[81] See the non-importation agreement, signed by Gloucester fish merchants Epes Sargent, Nathaniel Allen, Daniel Sargent, Winthrop Sargent, William Ellery Jr. dated Gloucester, May 2, 1768; the list of Boston merchants in violation of the non-importation agreement, dated Boston, April 27, 1769; and the memorandum of Joseph Pierce, dated Boston, June 10, 1769, Ezekiel Price Papers, 1754–1785, MHS.

[82] For more on the economic and political purposes of colonial non-importation agreements in general, see Marc Egnal and Joseph A. Ernst, "An Economic Interpretation of the American Revolution," *WMQ*, Vol. 29, No. 1 (January 1972), 21–23.

not show his face again in this port."[83] Similar tarring and feathering went on in other colonial ports.[84]

Colonial mobs also specifically targeted customs vessels in an effort to thwart and change the direction of imperial policy. The customs officers for Salem and Marblehead, for example, informed the Naval Board established in Boston to facilitate customs collection "of the loss of the Custom house Boat, which was maliciously and in the most secret and private manner set on fire about twelve o'Clock last night, and was entirely consumed by four this morning." The "Perpetrators of this act" took "the Boat from the Wharf to a considerable distance in the harbor." There, they "moored her with the Chain so that no assistance from the shore could possibly be given to save her."[85] In other colonial ports, mobs carried customs vessels to the commons, or in front of town meeting halls, before setting them on fire.

In sum, on both sides of the Atlantic there were political consequences associated with New England's maritime commercial expansion in the West Indies. British sugar interests perceived Yankee fish sold to French sugar producers with abhorrence, as such trade contributed to the ability of the French to seize Atlantic market share in the sugar trade. Business competition between British and French sugar suppliers led to the mobilization of political power. British sugar interests effectively used their seats in Parliament and their lobbyists to get protection from the British state in the form of various commercial regulations culminating in the Sugar Act. The Act adversely impacted the colonial fishing industry and stimulated certain legal and illegal forms of political protest in the colonies.

The Sugar Act, then, understandably frustrated Edward Payne and his fellow members of the Society for the Encouragement of Trade and Commerce. Colonial protests during the 1760s were forceful in word and deed. To be sure, colonial frustrations and protests did not immediately precipitate armed revolt against the British Empire. The American Revolution

[83] "Copy of Thomas Row's Affidavit concerning the insult he received from the Mob at Salem," dated Castle William, Boston Harbor, September 9, 1768; and "Copy of a Letter from the Collector & Comptroller of Salem and Marblehead to the Commissioners of the Customs at Boston," NA T 1/465/259–260, 265–266.

[84] Dirk Hoerder, *Crowd Action in Revolutionary Massachusetts, 1765–1780* (New York: Academic Press, 1977), 185–190.

[85] "Extract of a Letter from the Collector & Comptroller of Salem & Marblehead, dated the 20th June 1768," NA T 1/465/257.

did not begin directly after the Sugar Act, yet it had become clear that the British state consistently sided with sugar interests over colonial fishing interests. Those concerned in the colonial fishing industry believed sugar had corrupted their government. As a result, there was a popular belief in New England that the British state did not support the encouragement of colonial economic development. Additional events in northern Atlantic waters helped widen the gulf between these colonies and the metropole.

6

Atlantic Business Competition and the Political Economy of Cod: Part Two

> I now Enclose you four Depositions relative to the interruption our Fishery has met with on the coast of Labrador.[1]

Skipper Jonathan Millet and Company of the schooner *Hawke* were "a-Fishing in the Straits of Belle Isle near Bradore [Labrador]" on a summer fare in July 1766 when Captain Hamilton, "the Master of his Majesties' Sloop *Merlin* with a Number of Armed Men in several Boats came on board my Schooner, and demanded my Business there." The surprised yet pugnacious Yankee skipper "told him I was on a cod fishing Voyage from New England." The British naval officer treated Millet "as an Enemy... He damned me and gave a Blow with his Fist on my Face. I told him it was hard Usage. He again damned me & then ordered a Number of armed Men to seize me which they did. One of them at the same time shoving a Loaded Pistol against my Breast." Captain Hamilton informed the colonial fishing crew they "were a Pack of Damned Rascals several times." According to Millet, "He then seized my Vessel & Fish and put a Sentry on board." The captain of the HMS *Merlin* "kept [i.e. pressed] one of my Men, viz. Francis Con, & threatened that if he ever Catched any New England Men Fishing there again that he would seize their Vessels & Fish and keep all the Men, beside inflicting severe Corporal Punishment on every man he took."[2]

[1] Thomas Cushing, Speaker of the House of Representatives, in Boston, to Dennys DeBerdt, Massachusetts's Colonial Agent in London, Boston, January 17, 1767. Misc. Bound Manuscripts, 1766–1769, MHS. Labrador is located northwest of Newfoundland.

[2] Jonathan Millet, Deposition, September 13, 1766, Misc. Bound Manuscripts, 1766–1769, MHS. The Straits of Belle Isle divide the north coast of Newfoundland from the southern

Why would a British naval officer perform such actions against fellow subjects of the King of Britain? One explanation is that the officer was simply following orders. The Royal Navy had been directed to enforce the tenets of the 1763 peace treaty, which included keeping British fishermen out of fishing waters reserved for the French. And the Straight of Belle Isle *was* one of the regions the treaty made strictly off limits to British fishermen.[3] Indeed, when Millet demanded to know why the *Hawke*, her crew, and her catch were being seized, Captain Hamilton informed the skipper that he "had broke the Law."[4] However, maintaining peaceful international relations was not the only motivation behind the Royal Navy's vigilance and diligence in the North Atlantic. Maritime business competition played an important role in the Royal Navy's efforts to restrict New Englanders' access to fishing waters around Newfoundland.[5]

By the late eighteenth century, there were primarily three separate producers of dried, salted cod within the British Empire. Together, these producers brought £605,000 in annual export revenue into the British Empire between 1764 and 1775. This made dried, salted cod the fourth most valuable trade good in the Western Hemisphere for the British Empire.[6]

tip of Labrador. Salem fish merchant Jonathan Orne owned the *Hawke*. For a similar account of the same event, see Samuel Masury, Deposition, September 13, 1766, Misc. Bound Manuscripts, 1766–1769, MHS.

3 Article five of the 1763 Paris Peace Treaty stipulated that "the subjects of France shall have the liberty of fishing and drying on a part of the coasts of the island of Newfoundland, such as is specified in the XIIIth Article of the Treaty of Utrecht [i.e. from Bonavista to Point Riche]." Clive Parry, ed., *The Consolidated Treaty Series*, Vol. 42 (New York: Oceana Publications, Inc., 1969), 325. For article thirteen of the 1713 Utrecht Peace Treaty, see ibid., Vol. 27, 486.

4 When the British naval officer informed Millet that by international law the colonial fishermen were operating in illegal waters, the maritime laborer replied, "[I]f I had it was through Ignorance." Jonathan Millet, Deposition, September 13, 1766, Misc. Bound Manuscripts, 1766–1769, MHS. Lines drawn in peace treaties were not easy to see on the ocean. For more on this point, see Joshua M. Smith, *Borderland Smuggling: Patriots, Loyalists, and Illicit Trade in the Northeast, 1783–1820* (University Press of Florida, 2006).

5 As discussed in chapter five, the provisions of the 1764 Sugar Act and the commercial rivalry between British and French sugar interests were also responsible for the navy's efforts to control the movements of New England's commercial vessels in the North Atlantic. But, the Act only prevented Yankees from purchasing sugar products at St. Pierre and Miquelon. It did not explicitly pertain to commercial fishing activities in the North Atlantic. Therefore, the Act is not included here as a possible explanation for the treatment of Millet and his fishing crew.

6 Stephen J. Hornsby, *British Atlantic, American Frontier: Spaces of Power in Early Modern British America* (University Press of New England, 2005), 26–28, esp. figure 2.1.

The oldest and most politically connected cod fishery was headquartered in England's West Country. This region has had a long and well-documented relationship with the Atlantic cod fishing industry.[7] Ports such as Bristol, Plymouth, Barnstaple, Dartmouth, Poole, and Teignmouth were ideally located to prosecute the cod fisheries to the westward. On a small scale, fishermen from these coastal communities had long worked waters off Ireland and Iceland. Fish merchants in Bristol then financed John Cabot's westward explorations that resulted in the discovery of Newfoundland in 1497.[8] In 1501, these entrepreneurs formed a syndicate, "The Company Adventurers to the New Found Lands," to establish migratory commercial fishing operations off the coast of the land of cod.[9] Such expeditions entailed sending fishermen, vessels, and equipment during the summer months to catch and process cod for export to European markets. In 1578, there were an estimated fifty West Country fishing vessels working the waters around Newfoundland. By

[7] For a concise overview of the role West Country ports played in the medieval and early modern cod fishing industry, see Todd Gray and David J. Starkey, "The Distant-Water Fisheries of South West England in the Early Modern Period," in David J. Starkey, Chris Reid, and Neil Ashcroft, eds., *England's Sea Fisheries: The Commercial Sea Fisheries of England and Wales Since 1300* (London: Chatham Publishing, 2000), 96–104. There is a large literature on the role West Country merchants played in the cod fisheries around Newfoundland. For an overview, see Hornsby, *British Atlantic, American Frontier*, chapter 2. For the seventeenth century in particular, see Peter E. Pope, *Fish into Wine: The Newfoundland Plantation in the Seventeenth Century* (University of North Carolina Press, 2004), esp. 144–149. For the eighteenth and nineteenth centuries, see W. Gordon Handcock, *Soe longe as there comes noe women: Origins of English Settlement in Newfoundland* (St. Johns, NL: Breakwater Books, 1989), esp. 53–69 and table 3.1. For more on the role West Country fish merchants played in the settlement of New England, see Phyllis Whitman Hunter, *Purchasing Identity in the Atlantic World: Massachusetts Merchants, 1670–1780* (Cornell University Press, 2001), 33–70; David Sacks, *The Widening Gate: Bristol and the Atlantic Economy, 1450–1700* (University of California Press, 1991), 50, 101–103; and Bernard Bailyn, *The New England Merchants in the Seventeenth Century*, 3rd ed. (Harvard University Press, 1982).

[8] Sacks, *The Widening Gate*, 34–36. For more on speculations about pre-Cabotian Bristol voyages to Newfoundland, see Brian Fagan, *Fish on Friday: Feasting, Fasting and the Discovery of the New World* (New York: Basic Books, 2006), 202–205; and K. R. Andrews, *Trade, Plunder and Settlement: Maritime Enterprise and the Genesis of the British Empire, 1480–1630* (Cambridge University Press, 1984), 44–47. Mark Kurlansky feels very strongly that the Basques were working Newfoundland waters long before Cabot's voyage. Mark Kurlansky, *Cod: A Biography of the Fish That Changed the World* (New York: Alfred A. Knopf Canada, 1997), 17–29. Yet, in the most recent scholarly work on early Newfoundland, Peter Pope dismisses these early voyages, particularly the Basque ventures: "The notion of pre-Columbian Basque fishers in Newfoundland waters is a recurring fantasy of writers with more interest in a good story than in the evidence." Pope, *Fish into Wine*, 17, footnote 7.

[9] Sacks, *The Widening Gate*, 49–50; Harold Adams Innis, *The Codfisheries: The History of an International Economy* (Yale University Press, 1940), 12.

1592, this figure had doubled.[10] And by the late sixteenth and early seventeenth centuries, the West Country ports dominated the fishing industry in England.[11]

The youngest and the least politically powerful fishery operated out of Newfoundland.[12] Peter E. Pope maintains that it was not until 1677 that Newfoundland's resident fishing population became comparable to the resident stations on New England's coast.[13] As W. Gordon Handcock demonstrates, however, the number of resident fishermen in Newfoundland did not outnumber the migratory fishermen on the island until the 1790s. Moreover, the production levels associated with this nascent resident fishery were small in scale until the mid-to-late eighteenth century.[14]

Between these two poles sat the fishery headquartered in New England. West Country fish merchants financed the formation of resident fishing stations here during the first half of the seventeenth century. The English Civil War and London financiers helped colonists break away from the West Country's economic control. And by the eighteenth century, Yankees were building their own vessels and shipping cod to overseas markets themselves.[15]

These three cod fisheries competed with one another, despite the fact that each business was British. There were only so many fish in the sea and only so many consumers on land. Whoever controlled access to resources, produced the most product, and dominated markets, reaped the lion's share of the profits from the Atlantic cod trade.

Each of the three British fisheries competed for access to cod around Newfoundland. This competition was tacitly acknowledged in Parliament's 1699 "Act to Encourage the Trade to Newfoundland." Following pressure from fish merchants in England, the government made it a law that "no Alien or Stranger whatsoever (not Residing within the Kingdom of England, Dominion of Wales, or Town of Berwick upon Tweed) shall at any time hereafter take any Bait, or use any sort of Trade or Fishing Whatsoever in Newfoundland, or in any of the said Islands or Places

[10] Todd Gray, "Fisheries to the East and West," in "The Distant-Water Fisheries of South West England in the Early Modern Period," in Starkey, Reid, and Ashcroft, eds., *England's Sea Fisheries*, 97.

[11] Throughout the medieval period, England's eastern fishing ports led the English fishing industry. Wendy R. Childs and Maryanne Kowalski, "Fishing and Fisheries in the Middle Ages," in Starkey, Reid, and Ashcroft, eds., *England's Sea Fisheries*, 19–28.

[12] See the discussion in chapter six.

[13] Pope, *Fish into Wine*, 40.

[14] Handcock, *Soe longe as there comes noe women*, 74–85, esp. figure 4.2.

[15] See chapters two and four.

above mentioned." Defining "Aliens" and "Strangers" as all those "not Residing within the Kingdom of England, Dominion of Wales, or Town of Berwick upon Tweed" was clearly meant to reserve Newfoundland for fish merchants in England. The Act also established a system that gave special authority to fishermen from England. It stipulated that "every such Fishing Ship from England, Wales, or Berwick, or such Fishermen as shall from and after the said Twenty fifth Day of March, first enter any Harbor or Creek of Newfoundland, in behalf of his Ship, shall be Admiral of the said harbor or Creek during that Fishing Season, and for that time shall reserve to himself only so much Beach or Flakes, or both, as are needful for the Number of such Boats as he shall there use." It was observed that "several Inhabitants in Newfoundland" had since 1685 "Engrossed and Detained in their own hands, and for their own private Benefit, several Stages, Cook-Rooms, Beaches, and other Places in the said Harbors and Creeks (which before that time belonged to Fishing Ships [i.e. the migratory fishery]) for taking of Bait, and Fishing and Curing their Fish, to the great prejudice of the Fishing Ships that arrive there in the Fishing Season." Parliament therefore made it a law that no one should get these stages, cook rooms, etc., "before the Arrival of the Fishing Ships out of England, Wales and Berwick." The private property of West Country merchants was made inviolate. Their stages were not to be torn down; no one was to "Obliterate, Expunge, Cut out, Deface, or any Wise alter or Change" their "Mark or Marks" of ownership on vessels or equipment that were left until the following summer.[16] There would have been no need for this legislation had it not been for the business competition fish merchants in England faced at Newfoundland.

Each of the three British fisheries increased production of dried, salted cod over the course of the eighteenth century. The English Civil War only temporarily disrupted the West Country's ability to supply Iberian markets. In 1710, the British cod fishing interests at Newfoundland, understood as both the resident fishery and the West Country migratory fisheries, produced 100,000 quintals. Production then increased to over 300,000 quintals in 1730, and over 400,000 quintals in 1750. By the end of the century, in 1790, production levels at Newfoundland had reached 700,000 quintals.[17] Similarly, the New England cod fishing industry expanded its scale of operations over the course of the century following

[16] *An Act to Encourage the Trade to Newfoundland* (London: Printed by Charles Bill, 1699).

[17] Handcock, *Soe longe as there comes noe women*, 76, figure 4.1.

the Civil War.[18] By 1765, the fishing industry produced 350,500 quintals of dried, salted cod.[19] By comparison, Newfoundland cod fishing interests extracted 600,000 quintals in 1770. However, roughly an equal percentage of this catch was brought in by the West Country migratory sector of the Newfoundland fisheries and the permanent inhabitants living on the land of cod.[20] In other words, the New England resident fishing industry's output in 1765 was 350,500 quintals, while the West Country's migratory fishery produced 300,000 quintals in Newfoundland in 1770, and the resident fishery on the land of cod produced 300,000 quintals at the same time. It is likely, as we shall see, that New England production levels had dropped off by 1770. Yet, the Yankees' capability in 1765 to produce 350,500 quintals of dried, salted cod for export did pose a direct threat to West Country fish merchants, who had traditionally dominated the Atlantic cod trade.

Each of the three branches of the British Atlantic cod fishing industry also competed for control of the most lucrative market: the Iberian Peninsula. "The chief Branches of the British Fish Trade to Portugal," reported British consul William Poyntz in 1718, "consists in dry Cod from Newfoundland & New England; of which sort large quantities are expended [i.e. sold and consumed] yearly in this Kingdom."[21] Spanish and Portuguese consumers offered the best prices for dried cod. Iberian consumers paid on average twenty-seven to forty-four per cent more for merchantable-grade, dried, salted cod than West Indian consumers were willing to pay for refuse-grade cod throughout the colonial period.[22] Moreover, Iberian merchants were willing and able to pay for their fish in solid specie, while West Indians tended to exchange sugar products for fish. This, again, helps to explain why the principal economic function of the seventeenth-century colonial New England cod fisheries within the Atlantic economy was the transference of New World gold and silver into London coffers. Though this function shifted toward supplying West Indian slave plantations in the eighteenth century, the Yankees did not entirely give up on the more lucrative Iberian markets.[23] For their part,

[18] See the discussion in chapter four.

[19] Thomas Jefferson, Secretary of State, "Report on the State of the Cod Fisheries, 1791," *American State Papers: Commerce and Navigation*, I:13.

[20] Handcock, *Soe longe as there comes noe women*, 75–76, figure 4.1.

[21] William Poyntz to the Lords Commissioners of Trade and Plantations, Lisbon, September 17, 1718, NA SP 89/26/78A.

[22] This is discussed in chapter four.

[23] In 1793, despite the declension of the fisheries that resulted from the American Revolutionary War and the complete prohibition against the importation of American saltfish

West Country entrepreneurs perpetually dominated the British supply of Iberian markets, even when their position was weakened during the Civil War.[24]

The effectiveness of New England fish merchants in catching cod off the Newfoundland coast, processing more fish, and capturing market share among Iberian consumers frustrated West Country entrepreneurs. As early as 1715, fish merchants from Bideford and Poole petitioned Parliament for redress of the fact that New Englanders were catching cod off Newfoundland and thereby threatening the migratory fishery there.[25] But, there were additional reasons West Country merchants viewed New Englanders as a threat to their business interests.

By the middle of the eighteenth century, Yankee trade vessels were also transporting greater amounts of supplies, including lumber, fishing gear, and provisions of all sorts, to British settlers in Newfoundland. In exchange, the New Englanders received mainly dried cod, manufactured goods from Europe, and bills of exchange.[26] The number of New England trade vessels operating around Newfoundland increased from thirty-one in 1716 to 103 in 1751.[27] In part, this expanded trade resulted from

anywhere in the British Empire, which was our foremost trading partner at the time, dried, salted cod was the fifth leading export in the entire United States (behind breadstuffs, tobacco, rice, and wood, in that order). The industry had rebounded from the disruption of the Revolutionary War somewhat by the 1790s, and trade with France, Spain, and Portugal remained lucrative, although there was an import tax on American salted fish in place in France. See Thomas Jefferson, Secretary of State, "Report on the Privileges and Restrictions on the Commerce of the United States in Foreign Countries," December 16, 1793, *American State Papers: Foreign Relations*, I: 300–301.

[24] Ian K. Steele has pointed out that as early as 1698, sixty-four per cent of the West Country fishing fleet (N=135/210) traveled to ports in Spain and Portugal and then returned to England. Ian K. Steele, *The English Atlantic 1675–1740: An Exploration of Communication and Community* (Oxford University Press, 1986), 84. In addition, Portugal and Spain remained the largest consumers of "English cod" from Newfoundland well into the twentieth century. Álvaro Garrido, "Political Economy and International Trade: The Portuguese Market for Salt Cod and Its Institutions in the Interwar Period," *IJMH*, Vol. 17, No. 2 (December 2005), 61–85, esp. 67 and figure 1. Portugal's historic dependence on foreign cod, "the cod problem," had become so pronounced by the twentieth century that, on average, ninety per cent of the dried cod consumed in the region between 1900 and 1933 had been imported. Even as late as this modern period, much of this cod trade was controlled by businessmen in England. Ibid., 63, 69.

[25] C. L. Cutting, *Fish Saving: A History of Fish Processing from Ancient to Modern Times* (New York: Philosophical Society, 1956), 141.

[26] C. Grant Head, *Eighteenth Century Newfoundland* (Toronto: McClelland and Stewart Limited, 1976), 100–132; Ralph Greenlee Lounsbury, *The British Fishery at Newfoundland, 1634–1763* (Hamden, CT: Archon Books, 1969), 190–203; and Innis, *The Codfisheries*, 145, 146–147.

[27] Innis, *The Codfisheries*, 145.

increased permanent population levels at Newfoundland. Settlement at the land of cod expanded rapidly in the middle part of the century.[28] More people meant more mouths to feed, and Newfoundland's topography was not conducive to large-scale agriculture. As a result, and for reasons not totally dissimilar to the case of the West Indies, there was a heightened need for imported foodstuffs at Newfoundland. Shrewd Yankee traders sailed into the North Atlantic, as they had in the Caribbean, with full vessels to meet this growing demand for provisions.

West Country fish merchants opposed the New England provision trade with Newfoundland primarily for two reasons. It posed a threat to their own flow of supplies from England, and it provided an encouragement to permanent settlement on the island. Slowly but surely, New England entrepreneurs had been seizing market share in the provision trade to Newfoundland. By 1763, twelve per cent of the total quintals of cod caught at Conception Bay, Newfoundland, were being exchanged with colonial trade vessels for provisions. At the same time, thirty-two per cent of the quintals of cod caught at Ferryland and seven per cent of the quintals caught at Old Perlican and Trinity were exchanged thus. In all, 13,697 quintals were sold to colonial merchants, mostly from New England, in exchange for lumber and provisions.[29] In 1764, Commodore Hugh Palliser, the British naval officer and acting royal governor in charge of the island, calculated "A General Scheme of the Great Fishery." While he did not distinguish between fishing and trading vessels, nor did he provide the port of origin, he observed that there were 105 "Ships from America" of 6,337 tons burden operating around Newfoundland. Given New England's historic association with the fishing industry and trade with Newfoundland, it is almost certain that a large majority of these vessels were from New England ports. Moreover, the average of sixty tons per vessel fits well within the range for schooners, the Yankee vessel of choice in the eighteenth century. The largest number of "Ships from America," fifty-one, operated in the waters around St. John's. By comparison, there were 141 "British Fishing Ships," which included those belonging to settlers and "all ships arriving from England and Ireland whether for Fishing, with Supplies, or Passengers, or for Cargoes" of 14,819 tons burden. There were also ninety-seven "British Sack Ships," representing only West Country vessels "arriving from Foreign Ports,

[28] Handcock, *Soe longe as there comes noe women*, 73–120.
[29] Innis, *The Codfisheries*, 146.

whether with Salt or for Cargoes of Fish," of 11,924 tons burden.[30] Given the fact that in 1764 there was a total of 238 British vessels operating around Newfoundland, the principal locus of cod fishing operations in the eighteenth-century Atlantic world, the "American," or New England, presence was equal to forty-four per cent of this total. Or, to put this figure another way, New England's commercial activity around the land of cod represented almost half the level of activity attained by the three British Atlantic fishing interests.

New England's provision trade to Newfoundland provided an encouragement to the island's permanent settlement, which further enraged West Country merchants. Throughout the seventeenth and eighteenth centuries, West Country merchants fought permanent settlement at Newfoundland as it challenged their hegemony in the region, and they viewed anyone who supported such colonization with disdain.[31] In the first half of the seventeenth century, these merchants organized the Western Adventurers, a commercial consortium, to lobby Parliament against authorizing plans for permanent plantations at Newfoundland.[32] In 1668, they proposed forcibly removing settlers from the area.[33] At least in part, these entrepreneurs resented the fact that they had helped to finance settlement in New England, and subsequent generations of settlers had built a rival industry. West Country merchants were thus wary of plans to establish additional resident fishing stations in North America, particularly on Newfoundland. Yet, despite their best efforts, Newfoundland's permanent population slowly but surely increased.

[30] NA CO 194/16/109.

[31] There is a debate over the extent of conflict between English fish merchants and Newfoundland settlers. Gillian T. Cell, Ralph Lounsbury, Harold Innis, and Bernard Bailyn have all maintained that there existed very high levels of conflict between these two parties. See Gillian T. Cell, *English Enterprise in Newfoundland, 1577–1660* (University of Toronto Press, 1969); Lounsbury, *The British Fishery at Newfoundland*; Bailyn, *The New England Merchants in the Seventeenth Century*; and Innis, *The Codfisheries*. For their part, Peter Pope and Keith Matthews have argued that there was more symbiosis than friction between the merchants and settlers. See Pope, *Fish into Wine*; and Keith Matthews, "A History of the West of England – Newfoundland Fisheries" (Ph.D. Dissertation, Oxford University, 1968). W. Gordon Handcock and Grant Head can be pressed into this later crew, as they downplay the role that such conflict played in retarding permanent settlement at Newfoundland. Handcock, *Soe longe as there comes noe women*; and Head, *Eighteenth Century Newfoundland*. Despite this spirited and long-standing debate, none of these scholars disputes the fact that there was competition between the two groups.

[32] Hornsby, *British Atlantic, American Frontier*, 33.

[33] Lounsbury, *The British Fishery at Newfoundland*, 190.

As a result, savvy West Country merchants tacked a different course, and, by the turn of the eighteenth century, they focused their energies on stopping the flow of supplies from New England to Newfoundland. In 1696, West Country merchants complained that New Englanders enabled settlers to stay at Newfoundland by providing them with supplies.[34] In 1715, West Country merchants again petitioned Parliament for redress of their grievances, chiefly that Newfoundland "inhabitants are supplied with provisions, tobacco, rum, sugar, rice, etc., from New England and the colonies of America."[35] At this time, tobacco, sugar, and rice were enumerated articles, restricted by Navigation Acts and prohibited from direct trade with foreigners. But because there was no customs office in Newfoundland, New England trade vessels could vend these goods there in open marts to sack ships from Europe.[36] The West Country merchants, in short, tried to show that New England's trade violated Navigation Acts and must, therefore, be stopped. In another petition that same year, the West Country fish merchants referred to New England as "a nest of little peddlers," and they asked that direct trade between New England and Newfoundland be prohibited.[37]

Finally, New Englanders transported migratory fishermen as passengers away from Newfoundland, which further angered merchants in England's West Country.[38] At the end of the seventeenth century, the West Country fish merchants complained: "[T]he worst thing is that the New England men carry away many of the fishermen and seamen, who marry in New England and make it their home."[39] By these means, the migratory fishing industry suffered a loss of manpower while the colonial industry gained. According to an estimate made between 1713 and 1718, 1,000 migratory fish servants booked passage on Yankee vessels bound for New England ports every year.[40] In 1764, Commodore Palliser observed "a constant current of seamen, artificers and fishermen through this country into America."[41] A report in 1771 stated that "New England vessels

[34] Ibid., 193.

[35] NA CO 194/5. Cited in Innis, *The Codfisheries*, 144.

[36] Head, *Eighteenth Century Newfoundland*, 111–112.

[37] NA CO 194/7. Cited in Innis, *The Codfisheries*, 145.

[38] Hornsby, *British Atlantic, American Frontier*, 33; Handcock, *Soe longe as there comes noe women*, 29–30; Lounsbury, *The British Fishery at Newfoundland*, 194, 223–224; and Bailyn, *The New England Merchants in the Seventeenth Century*, 130.

[39] Cited in Innis, *The Codfisheries*, 103.

[40] Handcock, *Soe longe as there comes noe women*, 29–30; and Innis, *The Codfisheries*, 147. Also see Pope, *Fish into Wine*, 243–248.

[41] Cited in Innis, *The Codfisheries*, 147.

last year carried out of Conception Bay [Newfoundland] upwards of 500 men, some of which were headed up in casks because they should not be discovered."[42] Push factors such as harsh working conditions and low wages motivated eighteenth-century workers to leave the land of cod.[43] Higher wages, more women, and more temperate climes pulled these men into New England. One Newfoundlander wrote in 1700, "[T]here being great wages given to men in New England makes men desirous to go there, and frequently attempts it."[44] For Yankee merchants, passenger fees increased profit margins by augmenting earnings from trade missions and fishing expeditions to Newfoundland.

To slow down, stop, and even reverse New England's maritime commercial expansion, West Country fish merchants sought, and gained, government protection. The West Country entrepreneurs successfully persuaded the British government to protect their commercial interests for several reasons. West Country commercial fishing interests played a significant role in early modern England's economic development. Merchants from this region provided much of the capital behind England's westward expansion into the Atlantic.[45] England's chief wool producers also resided in the West Country.[46] Moreover, the eighteenth century marked the heyday of the English migratory cod fisheries around Newfoundland. During this period, cod caught off Newfoundland and sold in Southern European markets represented one of British North America's strongest export revenue streams.[47]

West Country merchants also had influential political connections. Like West Indian planters, they tended to be absentee landlords. The wealth generated through the Newfoundland cod fisheries, like sugar capital, enabled a genteel lifestyle in England.[48] Just as their tropical counterparts had done, the West Country merchants used their wealth

[42] Cited in Innis, *The Codfisheries*, 104. The date is provided by Handcock, *Soe longe as there comes noe women*, 29, footnote 23.
[43] Handcock, *Soe longe as there comes noe women*, 85.
[44] Cited in Daniel Vickers, *Farmers and Fishermen: Two Centuries of Work in Essex County, Massachusetts, 1630–1830* (University of North Carolina Press, 1994), 109. Also, see Pope, *Fish into Wine*, 159–160, 183. Of course, not every worker who left Newfoundland ended up financially better off. In 1763, Thomas Clark, an Irish fishing servant, fled Newfoundland only to end up "representing his distressed Circumstances, and praying Relief" from the Massachusetts General Court in Boston. *JHRM*, Vol. 39, 146.
[45] See chapter three.
[46] Hornsby, *British Atlantic, American Frontier*, 10.
[47] Ibid., 26–28, esp. figure 2.1.
[48] Ibid., 38, 40–41, figure 2.8.

and land ownership in England to control seats in Parliament.[49] Benjamin Lester, for example, was one of the wealthiest fish merchants in Poole, England. He controlled a resident fishing station in Trinity, Newfoundland, with its own wharf space, warehouse, shop, flakeyard, shipyard, sail loft, forge, cookroom, and vessels. Lester also owned real estate in Poole, served as the town mayor, and as a member of the House of Commons. In his mansion house in Poole, Lester had two dried cod fillets carved in marble on his fireplace mantelpiece to symbolize the source of his wealth and political power.[50] They also hired professional lobbyists to protect their interests throughout the colonial era.[51]

Moreover, Parliament responded favorably to the claim of West Country merchants that the migratory fishing industry represented the authentic, traditional British "nursery" for naval seamen. This role had been codified into law with the 1699 Act, which remained in effect until 1867.[52] Leading lights such as Adam Smith equated "the extension of the fisheries of our colonies" with an "increase [in] the shipping and naval power of Great Britain."[53] The assumption here was that commercial fishing trained men to handle wooden sailing vessels, gave them knowledge of

49 For more on the involvement of West Country fish merchants in Parliament during the seventeenth century, see Lounsbury, *The British Fishery at Newfoundland*, 210, 213. For the eighteenth century, see Hornsby, *British Atlantic, American Frontier*, 30.

50 Hornsby, *British Atlantic, American Frontier*, 30, 34, 38–39, 40–41, figures 2.5, 2.7, 2.8; and Jerry Bannister, *The Rule of the Admirals: Law, Custom, and Naval Government in Newfoundland, 1699–1832* (University of Toronto Press; for the Osgoode Society for Canadian Legal History, 2003), 156–158.

51 Pope, *Fish into Wine*, 188; and Hornsby, *British Atlantic, American Frontier*, 33.

52 Migratory fishing vessels leaving England were, by law, supposed to "Carry with them in their Ships Company, at least one such Fresh Man that never was at Sea before, in every Five Men they Carry." This "Fresh Man" is also referred to as a "Green-Man (that is to say) not a Seaman, or having been ever at Sea before." In this manner, men would be trained to life and work at sea, which would establish a labor pool that the navy could then draw upon. *An Act to Encourage the Trade to Newfoundland* (London: Printed by Charles Bill, 1699).

53 Adam Smith, *An Inquiry into the Nature and Causes of the Wealth of Nations*, Max Lerner, ed. (New York: The Modern Library, 1937; orig, pub. in 1776), 544. For an early seventeenth-century argument that the way to increase maritime commerce and naval strength was to follow the Dutch model and encourage a commercial fishing industry, see Tobias Gentleman, *England's Way to Win Wealth, and to Employ Ships and Mariners* (London, 1614). The anonymous author of a 1653 pamphlet wrote that promoting commercial fishing "will maintain thousands of this Nation, increase Mariners for Navigation [maritime commerce], and greatly strengthen the Navy of this *Commonwealth*, and enrich the Nation, advance the Customs [tax revenue], and raise a great and yearly Revenue unto this *Commonwealth*." Expanding the fishing industry would further safeguard England's maritime sovereignty: "[T]his being effected, we shall be most absolute and powerful at Sea." Anonymous, *The Seas Magazine Opened* (London: 1653).

trade winds and ocean currents, and familiarized them with life at sea. Such training, it was believed, made commercial fishing ports ideal labor pools from which to draw manpower for the navy in wartime. Moreover, the fishing industry stimulated growth in shipbuilding and other maritime trades that could support naval power during periods of conflict. The expansion of the Royal Navy around the turn of the eighteenth century, the near-constant wars with France, and the huge demands for maritime labor that resulted, only served to deepen the importance of training English sailors.[54] As a result, eighteenth-century Parliamentarians were very sympathetic toward the migratory fishing industry.

Pressure from West Country fish merchants bore fruit in 1764. British naval officers had governed Newfoundland through the summer fishing season as a result of the 1699 Act.[55] Commodore Hugh Palliser ruled the region with an iron fist from 1764 to 1769. Heads of the British state instructed him to enforce the tenets of the 1763 peace treaty; to enforce commercial legislation such as the Sugar Act; and to actively support West Country fishing interests by more vigorously enforcing the 1699 Act.[56] Palliser, in turn, instructed the captains under his command to keep all British fishing vessels out of French territorial waters, and to prevent illicit trade with the French islands in the North Atlantic.[57] He also issued orders during his first year as governor expressly forbidding anyone to transport passengers from Newfoundland to the North American colonies without first entering on board his personal flagship and asking for his written permission. Violators risked having their vessels impounded and paying a stiff penalty.[58] In 1765, George III personally made it explicitly

[54] For more on the turn-of-the-century naval expansion and the demands for maritime labor, see N.A.M. Rodger, *The Command of the Ocean: A Naval History of Britain, 1649–1815* (New York: Norton, 2005); and Daniel Baugh, *British Naval Administration in the Age of Walpole* (Princeton University Press, 1965).

[55] Bannister, *The Rule of the Admirals*, 27, 148–149, 158–162.

[56] See Palliser's commission as governor, which was read in the fort at St. John's on June 19, 1764, Colonial Secretary Letterbooks, 1752–1765, MUMHA, 218. In a personal letter from Lord Egremont to Palliser, dated Whitehall, July 9, 1763, the British naval officer was formally warned to enforce the peace treaty: "[T]he King will not Pass over unnoticed any Negligence or Relaxations on the Part of any Persons employed in his Service in a matter on which His Majesty lays so much Stress and in which the fair Trade of all His faithful subjects is so Essentially interested." Ibid., 225.

[57] See the discussion in chapter six.

[58] See Palliser's Orders "for Vessels Trading to America not to carry Passengers & to sail by the 5th of next Month," St. Johns, October 4, 1764; Palliser's "Order to American Vessels not to convey Passengers," St. Johns, November 1, 1764; and Palliser's "Decree upon Thomas Stout, Master of the Brig *Good Intent*," St. Johns, September 16, 1765, Colonial Secretary Letterbooks, 1752–1765, MUMHA, 251–252, 278, 340.

clear that Palliser was to prevent "the great Prejudice & discouragement of the Ship Fishery," or the West Country migratory fishing industry.[59]

Eager to court royal favor, the surest route to promotion in the navy at this time, Palliser established what he believed to be effective measures for encouraging the "Ship Fishery." He initiated a program for forcibly removing New England fishing and trading vessels from Newfoundland waters. He ordered the warships under his command patrolling Newfoundland waters to confiscate any fish found on New England vessels operating in the region. Palliser further authorized the seizure of these vessels, and the use of corporal punishment on repeat Yankee offenders.[60] Thus, the 1763 peace treaty was not the sole source of Millet's treatment described in the opening of this chapter. The New England fish merchants complained of Palliser's actions through their agent in London.[61]

In response to the colonial entrepreneurs, the Board of Trade merely held a formal inquest, which was nothing more than a political show to present a façade of objectivity. Palliser's own words in defense of his actions are instructive on this point. In language the shrewd commander knew would clear him of any blame, he justified his stringent measures by stating that he wanted Newfoundland and the surrounding area to be a British, "not American," fishery. He added that allowing the "Americans" (by which he meant colonists in general, but New Englanders in particular) to fish in Newfoundland waters was "ruinous" to the "British" fishing industry.[62] In short, Palliser, like the West Country fishing interests and George III, saw the colonists not as fellow British subjects entitled to equal rights, but as mere interlopers in commercial activities belonging to entrepreneurs in England.[63] And the Board of Trade concurred. In the end, the Board favored the West Country interests and formally approved Palliser's actions by not censuring the aggressive commander.

[59] "George III's instructions to his royal governor, Hugh Palliser," St. James, May 10, 1765, Colonial Secretary Letterbooks, 1752–1765, MUMHA, 285.

[60] See the Ezekiel Price Papers, 1754–1785, MHS; and Letter from Gov. Palliser to Gov. Barnard at Boston, "Great St. Lawrence Harbor," July 15, 1765, Colonial Secretary Letterbooks, 1752–1765, MUMHA, 290. Also, see Jonathan Millet, deposition, September 13, 1766; and Samuel Masury, deposition, September 13, 1766, Misc. Bound Manuscripts, 1766–1769, MHS.

[61] See Dennys DeBerdt, Massachusetts's Colonial Agent, in London to the Boston committee for the encouragement of trade and commerce, London, March 14, 1767, Ezekiel Price Papers, 1754–1785, MHS.

[62] Palliser to the Board of Trade & Plantations in London, St. John's, Newfoundland, October 30, 1765, NA CO 194/16/170-174.

[63] Indeed, Palliser went so far as to publicly accuse New England fishermen of "piracy." *JHRM*, Vol. 43, Part II, 257.

Having secured the political support of the British state, and the military assistance of the Royal Navy, the West Country fish merchants must have been optimistic that the New England fishing industry would soon go into decline.

To be sure, Palliser's actions did have a cumulative negative effect on colonial New England's maritime commerce. The number of "Ships from America" operating around Newfoundland fell off from 105 in 1764 to 83 in 1766.[64] But, the stubborn colonists held fast and weathered the storm. The final hammer blow did not come until 1775.

In sum, there were political consequences associated with New England's maritime commercial expansion in the North Atlantic. Fish merchants headquartered in England's West Country viewed the fishing waters around Newfoundland as their private preserve. It did not sit well with them that colonial entrepreneurs had built a rival industry by trading and fishing at the land of cod. From 1699 to 1764, fish merchants in England successfully obtained varying levels of government protection, which helped them maintain their dominant position in the Atlantic cod trade. Colonists protested these events, but to no avail.

The incidents in the North Atlantic that troubled skipper Jonathan Millet and increased operating costs associated with the colonial fishing industry did not immediately lead to demands for independence among the colonists. Instead, the cumulative effect of efforts to restrict and regulate New England's commercial expansion along the Northwest Atlantic littoral contributed to a rising conviction among colonists that the British state did not support, and even actively opposed, their right to use the sea for commercial purposes. Parliament could have avoided taking an official position in the transatlantic business rivalry that existed between colonists and West Country entrepreneurs. An objective political posture would have weakened colonial resolve when it came to considering whether the British state was openly hostile to colonial maritime interests. This may, in fact, have been the initial intent of the official inquest into Palliser's actions. Instead, as the following chapter demonstrates, the government's position only hardened over time.

[64] NA CO 194/16/109, 194/16/317.

7

The New England Trade and Fisheries Act

> The Fishery on the Banks of Newfoundland, and the other Banks, and all the others in America, was the undoubted right of Great Britain. Therefore we might dispose of them as we pleased.[1]

On February 10, 1775, Prime Minister and Chancellor of the Exchequer Frederick North "moved for leave" in Parliament to introduce "a Bill for putting the trade of America with England, Ireland, and the West Indies under temporary restrictions, and for restraining the refractory provinces from fishing on the banks of Newfoundland."[2] Dissenting heads of the British state who were among the first to read copies of the bill were very concerned. In the House of Commons, Sir John Griffin "after expressing his sincere wishes to see a happy conclusion put to the American disputes without bloodshed, declared, that upon reading the Bill, he felt himself alarmed, and was jealous that, if the greatest caution and delicacy was not to be used in perfecting the Bill, it would rather provoke than effect any good purpose."[3] When the bill made it to the House of Lords, peers such as Charles Prat, Earl of Camden, a Whig and the judge who acquitted John Wilkes, compared the Bill to other Parliamentary legislation aimed at America. Lord Camden found that other laws were "by no means so violent in their operations as this." He believed that the bill "was at once declaring war [against the colonies], and beginning hostilities in Great

[1] Lord North's speech in Parliament dated February 10, 1775. *PDBP*, Vol. 5, 412; and *AA*, Series 4, Vol. 1, 1621–1626.

[2] *PDBP*, Vol. 5, 410. The official title of the bill at this point was the "New England Trade and Fishery Prohibitory Bill."

[3] *PDBP*, Vol. 5, 460.

Britain [i.e. the British Empire]."[4] The bill passed into law despite such objections. It became the New England Trade and Fisheries Act, commonly referred to as the Restraining Act.

This chapter will explain why the Act generated alarm on both sides of the Atlantic. The Act's origins will be established. The Act itself will also be detailed. Finally, its effects will be enumerated.

Lord North introduced the aforementioned bill into Parliament for several reasons. New Englanders did not have the best reputation in Whitehall during the late 1760s and early 1770s. The Stamp Act riots in New England, the tarring and feathering of customs agents who worked for the Treasury Department, the Boston Massacre, and the Boston Tea Party, collectively made it appear that the region was lawless and in a state of rebellion.[5] Moreover, the non-importation agreement the Boston-based Society for the Encouragement of Trade and Commerce had established and the Continental Association seemed to North and other MPs to circumvent the British government's right to regulate colonial commerce.[6] As early as 1766, Lord Grenville, who headed the Treasury Department before North, declared the colonists' behavior "highly criminal."[7] And during the same month that he introduced the aforementioned bill, Lord North, who had served as a junior member of the Treasury Board since 1759, and who had become Chancellor of the Exchequer in 1767, stated that New Englanders were "in a state of actual rebellion."[8] The bill must be seen then, at least in part, as a disciplinary effort to punish recalcitrant colonists who had stubbornly refused to submit to the doctrine of Parliamentary sovereignty when it came to issues such as taxation and commercial regulation. Indeed, North promoted his motion in part by declaring "that as the Americans had refused to trade with this Kingdom, it was but just that we should not suffer them to trade with any other

4 *Lord Camden's Speech on the New-England Fishing Bill* (Newport, RI: S. Southwick, 1775), Early American Imprints, Series I: Evans #42788.

5 For a standard overview of these events, see Robert Middlekauff, *The Glorious Cause: The American Revolution, 1763–1789*, rev. and expanded ed. (Oxford University Press, 2005).

6 The Continental Association was established at the end of 1774. It was a collective non-importation and non-exportation agreement that prohibited member American colonies from receiving goods from areas Great Britain controlled after December 1, 1774. The Association further prevented members from shipping provisions to these areas after September 10, 1775. *JCC*, Vol. 1, 75–80; and *NDAR*, Vol. 1, 16n.

7 *PDBP*, Vol. 2, 149.

8 *PDBP*, Vol. 5, 352.

Nation."[9] But, the bill was not solely, or even most prominently, a form of punishment and a reassertion of Parliamentary sovereignty.

The bill must also be seen as a Navigation Act. It was a mercantilist attempt to regulate colonial commerce in such a way as to make colonial economic interests subordinate to similar interests in the mother country.[10] To mercantilist members of Parliament who understood the financial importance of the British West Indian sugar islands and the West Country migratory fishing industry to the wealth of the mother country, it seemed as though pecuniary-minded New Englanders had been damaging imperial economic interests throughout the eighteenth century. These colonists had continued to sell fish to the French in the West Indies. They then smuggled cheap French sugar products back into the colonies and avoided customs duties.[11] North's bill was clearly meant to prevent Yankees from provisioning French sugar plantations, thereby squeezing a major source of smuggling. Closing the fishing industry eliminated New England's foremost export and severely reduced the profitability of sending ships into French territories.

Long-standing grievances stemming from business competition in the Atlantic cod trade further influenced the bill's introduction into Parliament. West Country fish merchants did not appreciate New England entrepreneurs poaching in their private preserve, taking cod from the waters around Newfoundland. The fish merchants in England also resented Yankee merchants provisioning residents at Newfoundland, which hurt the West Country supply business. Finally, the West Country merchants did not approve of New England shippers taking workers in the migratory fishing industry as passengers away from Newfoundland.[12]

West Country fish merchants actively supported North's bill, and they played on mercantilist heart strings in Parliament. Merchants from Poole, the foremost port in England involved in the migratory fishing industry at Newfoundland, sent a petition to be read before the members of Parliament while the bill was being debated in the opening months of 1775.[13]

[9] *AA*, Series 4, Vol. 1, 1622.

[10] For more on British mercantilist ideas and policies, see Cathy Matson, *Merchants & Empire: Trading in Colonial New York* (The Johns Hopkins University Press, 1998).

[11] See John W. Tyler, *Smugglers & Patriots: Boston Merchants and the Advent of the American Revolution* (Northeastern University Press, 1986); and William T. Baxter, *The House of Hancock: Business in Boston, 1724–1775* (Harvard University Press, 1945).

[12] See chapter six.

[13] Following passage of the bill, a reporter noted that certain businessmen in London "curse the Poolemen, for they say it was owing to them that the restraining act took place." *AA*, Series 4, Vol. 3, 256.

Their petition was meant in part to negate the arguments of London manufacturers who maintained commercial relations with merchants in New England, and who had testified on colonists' behalf.[14] MPs would have heard the West Country's voice loud and clear as the petition was read: "[T]he Petitioners beg Leave to observe, that the Restraints intended to be laid upon the Newfoundland Fishery of the Colonies, mentioned in the said Bill, if carried into a Law, will not by any Means be injurious to Commerce... because the Foreign Markets can be amply supplied by extending the Newfoundland Fishery *of Subjects resident in England*." The West Country merchants went on to explain "that the Annual Produce of the Newfoundland Fishery carried on *by Subjects resident in the Mother Country* exceeds £500,000." And they reminded Parliament "that the Profits of the Trade *center entirely in this Kingdom*," while in sharp contrast "the Profits of the Newfoundland Fishery carried on by the Colonies mentioned in the Bill *do not center here*."[15]

The West Country merchants then reminded MPs of the service the migratory fishery provided to Britain's navy. In words that did not fall on deaf ears among residents of an island nation, they stated that "the Newfoundland Fishery of the Mother Country is a constant Nursery of Seamen for the Navy, that great Bulwark of the Nation, every Fifth Man employed being, by the 10th of William the Third [the 1699 Newfoundland Act], obliged to be a Landman, a Consideration of infinite Weight." By contrast, "the Newfoundland Fishery of the Colonies [was not] a Nursery of Seamen for the Fleet, because the Americans are not obliged by Law to make use of Landman, nor are the American Seamen compellable [because of the "Sixth of Anne"], like the British Seamen, to serve their Country in Times of War." The merchants concluded by admitting that the bill would serve "their own immediate [economic] Advantage," but

[14] The London entrepreneurs were concerned that any threat to New England's foremost export, fish, would hinder colonists' ability to pay for manufactured goods from England. The Londoners argued "[t]hat the said Bill, as the Petitioners conceive, is unjustly founded, because it involves the whole, in the Punishment intended for the supposed Offenses of a few; and that it must, in its Consequences, overwhelm Thousands of His Majesty's loyal and useful Subjects with the utmost Poverty and Distress, in as much as they will be thereby deprived of the Fisheries which are the natural Means of supporting themselves and Families; and that the extensive Commerce between Great Britain and her Colonies will by this Bill be greatly injured, as a capital Source of Remittance will be stopped, which will not only disconnect the future Commercial Intercourse between those Colonies and this Country, but it will eventually render them incapable of paying the large Debts already due to the Merchants of this City." The petition was read twice in the House of Commons, once on February 23, 1775, and again the following day. *PDBP*, Vol. 5, 455, 459; and *AA*, Series 4, Vol. 1, 1633.

[15] *PDBP*, Vol. 5, 480. Emphasis in this section my own.

they felt it would also be "for the universal Benefit of their Country."[16] In short, closing the colonial cod fisheries would considerably reduce business competition for West Country interests, eliminate a threat to mercantilist policies, and encourage that great "nursery of Seamen for the Fleet."

In addition to this Poole petition, Parliament heard personal testimonies from those who supported the bill. Most witnesses with first-hand knowledge of the Atlantic cod trade were asked the following question repeatedly: "Whether there is not an established Fishery on the banks of Newfoundland from the West of England; and if the Fisheries of New England were stopped, they would not increase to supply its deficiencies?"[17] Members of Parliament wanted reassurance that the British Atlantic cod trade would not suffer too badly after commercial fishing was removed from New England. Parliamentarians also specifically wanted confirmation that the fishing industry in the mother country would benefit financially from this move.

Benjamin Lester, mentioned in the last chapter, testified that as a fish merchant from Poole he had been engaged in the Newfoundland fishery for "thirty-eight years." Such long-term familiarity with the business, combined with his prior political experience, which included holding a seat in the House of Commons, must have lent a particular weight to Lester's words. When asked, "Can the foreign markets be supplied, if the New England Fishery is stopped?" Lester's confident response without doubt impressed many heads of state: "They certainly may."[18] He reminded Parliamentarians that his homeport, along with Dartmouth, represented "the principal Ports in England from whence the Newfoundland Fisheries are carried on." Lester then acknowledged he and his fellow merchants stood to financially gain from the Newfoundland fishery being "confined to Great Britain only." He unequivocally stated that "if New England was restrained forever from this Fishery it would be a benefit to Great Britain." Lester explained "that if the American Fishery was stopped, other places in Great Britain besides Poole and Dartmouth would engage in it." What is more, the expansion of the fishing industry in England would "consequently increase the number of British Seamen" available for service in the navy.[19]

[16] *PDBP*, Vol. 5, 480. For more on the 1699 Act, see the discussion in chapter six. For more on the "Sixth of Anne," see chapter five, footnote 40.

[17] *AA*, Series 4, Vol. 1, 1638–1650. Versions of this question can be found throughout the depositions in this document. The quote is taken specifically from ibid., 1644.

[18] *AA*, Series 4, Vol. 1, 1651. Lester's last name is listed in this record as "Lister."

[19] *AA*, Series 4, Vol. 1, 1668. Here, Lester is listed as "Lyster."

Other persuasive testimonies contributed to the termination of New England's cod fisheries. George Davis, another merchant long-experienced with the migratory fishery at Newfoundland, testified "that of late years the *New England* Fishery is much increased, and the *British* Fishery very much decreased." Like Lester, Davis confirmed "that if this Act should pass," he would "reap benefit from it." Like Lester, Davis swore on oath "that the foreign markets might be supplied entirely from *Great Britain*, if the *New England* Fishery was stopped."[20]

Captain Molyneux Shuldham, head of the British Naval squadron stationed at Newfoundland and acting governor of the land of cod, was next to take the stand. He testified that the navy "cannot take any Seamen out of the New England Ships [because of the "Sixth of Anne"], but that a great many are got out of the British Fishing Ships [because of the 1699 Act]." Shuldham went on to state:

> that the New England Ships carry on an illicit trade with the French; that they load with Provisions and Lumber, and go to meet the French Ships at Sea; that they sell them Ship and Cargo, and take French Manufactures and India Goods in exchange; that the New England Ships carry Provisions to the French at Miquelon and St. Pierre.... that they supply the French Fishermen with Flour from Indian Corn; [and] that numbers of our Seamen desert [to] the New England Ships.[21]

Similar to Lester and Davis, the British Naval officer testified that "if this temporary restraint on the *New England* Fishery was made perpetual, it would be a benefit to *Great Britain*."[22]

The most damning testimony came last. Hugh Palliser, now a Baronet, was called and sworn in. He testified "that this [i.e. the West Country migratory] Fishery is the best nursery [for naval seamen]; that the men are better for the Men-of-War than those taken out of the Colliers; that it would be impossible to man a Fleet, but in a great while, if it was not for the men they get from the Newfoundland Fishery." This was a naval officer and an aristocrat testifying before Parliament on oath that the Royal Navy could not be sufficiently manned without the migratory fishing industry operating out of the West Country. Palliser corroborated Shuldham's testimony "that the New England Ships carry Provisions to [the French islands of] St. Pierre and Miquelon." Moreover, Palliser stressed "that whether the restraining of the New England Fishery is temporary or perpetual, it will be an advantage to Great Britain; that the Fishery

[20] *AA*, Series 4, Vol. 1, 1669. Emphasis in the original.

[21] Ibid.

[22] *AA*, Series 4, Vol. 1, 1669–1670. Emphasis in the original.

might be carried on from Great Britain, Ireland, Jersey, and Guernsey, which would greatly increase the nursery for Seamen."[23]

For all of these reasons, Lord North fought hard in Parliament to get the New England Trade and Fishery Prohibitory Bill passed.[24] The bill was initially meant to do two things: restrict the trade of the New England colonies "to Great Britain, Ireland, and the British Islands in the West Indies," and "prohibit such Provinces and Colonies from carrying on any Fishery on the Banks of Newfoundland." North wanted Yankee fish merchants to stop trading with the French sugar islands, *and* he wanted them to cease fishing operations in West Country waters in the North Atlantic. The law was to stay in effect, in North's mercantilist words, "till the New Englanders should return to a sense of their duty to the mother country, and submit to her supreme authority." In North's mind, "the Fishery on the Banks of Newfoundland, and the other Banks, and all the others in America, was the undoubted right of Great Britain. Therefore we might dispose of them as we pleased."[25] North believed, as did the West Country merchants, that rich oceanic resources should *not* be "disposed of" in favor of the New England colonies. Rather, North Atlantic fishing waters belonged to economic interests residing in the mother country. Indeed, North proposed "that an augmentation of 2,000 seamen should be made," and "a proper number of frigates" should be stationed "in the most beneficial manner for commanding the whole coast of North America." The sole justification for the expense associated with this naval augmentation was that without proper military force "the people of New England could not be restrained from the fishery."[26]

North had to go to some lengths to justify punitively impacting the entire New England region. Opponents demanded to know the following: "[W]hy was the whole to be punished? Why New-Hampshire? Why Rhode Island? Why Connecticut?"[27] North acknowledged that "the

[23] *AA*, Series 4, Vol. 1, 1670.

[24] North addressed critics of the bill in both the House of Commons and the House of Lords between February 10 and March 31. *PDBP*, Vol. 5, 480, 653. Over the course of these two months, critiques of North's bill and North's rebuttals take up more space in the records of Parliamentary debates than any other issue. The bill was of prime importance.

[25] *PDBP*, Vol. 5, 410, 411, 412. Protocol was established to evaluate precisely when colonists had returned to "a sense of their duty." For example, North allowed colonists to apply for fishing licenses only after having first submitted evidence of "their good behaviour" to royal governors "or upon their taking a test of acknowledgement of the rights of Parliament." Ibid., 413.

[26] *PDBP*, Vol. 5, 411, 418.

[27] *AA*, Series 4, Vol. 1, 1623.

two Houses [in Parliament] had not declared all Massachusetts Bay in rebellion," let alone all of New England. He reasoned, however, there were "armed mobs" throughout Massachusetts and in New Hampshire, Connecticut, and Rhode Island that posed a clear and present danger to Parliamentary sovereignty. Moreover, "the vicinity" of these colonies "to Massachusetts Bay was such" that if they were "not added, the purpose of the Act would be defeated." The "vicinity" of neighboring ports, North maintained, "afforded the inhabitants" of Massachusetts "an opportunity of carrying on the Fishery."[28] Thus, all of New England had to suffer.

Parliament voted the Restraining Act into law on March 21, 1775.[29] George III gave his royal assent nine days later. The Act arrived in Boston by April 27.[30] It was published in colonial newspapers as early as May 8.[31] The section pertaining to the fisheries formally stated:

> And it is hereby further enacted by the authority aforesaid, That if any Ship or Vessel, being the property of the subjects of Great Britain, not belonging to, and fitted out from Great Britain or Ireland, or the Islands of Guernsey, Jersey, Sark, Alderney, or Man, shall be found, after the twentieth day of July, one thousand seven hundred and seventy-five, carrying on any Fishery, of what nature or kind soever, upon the banks of Newfoundland, the Coast of Labrador, or within the River or Gulf of St. Lawrence, or upon the Coast of Cape Breton, or Nova Scotia, or any other part of the Coast of North America, or having on board materials for carrying on any such Fishery, every such Ship or Vessel, with her Guns, Ammunition, Tackle, Apparel, and Furniture, together with the Fish, if any shall be found on board, shall be forfeited, unless the Master, or other person having the charge of such Ship or Vessel, do produce to the Commander of any of his Majesty's Ships-of-War, stationed for the protection and superintendence of the British Fisheries in America, a Certificate, under the hand and seal of the Governor or Commander-in-Chief, of any of the Colonies or Plantations of Quebec, Newfoundland, St. John, Nova Scotia, New-York, New-Jersey, Pennsylvania, Maryland, Virginia, North Carolina, South Carolina, Georgia, East Florida, West Florida, Bahamas, and Bermudas, setting forth that such Ship or Vessel, expressing her name, and the name of her Master, and describing her build and burthen, hath fitted and cleared out from some one of the said Colonies or Plantations, in order to proceed upon the said

[28] *AA*, Series 4, Vol. 1, 1622–1623.

[29] *PDBP*, Vol. 5, 584; and *AA*, Series 4, Vol. 1, 1691–1696.

[30] On this date, Jebediah Huntington wrote from Cambridge, Massachusetts, to Jonathan Trumbull Jr., "Mr. Josiah Quincy is arrived from London, in a very low state of health, and not expected to live. The Restraining Act is come by the same ship." *AA*, Series 4, Vol. 2, 424.

[31] *Newport Mercury*, May 8, 1775; and *Virginia Gazette*, May 18, 1775.

> Fishery, and that she actually and bona fide belongs to, and is the whole and entire property of his Majesty's subjects, inhabitants of the said Colony or Plantation; which Certificates such Governors or Commanders-in-Chief, respectively, are hereby authorized and required to grant.[32]

The official Act that was made the law of the British Atlantic made it illegal for vessels from anywhere in New England to fish everywhere on "the Coast of North America." The Grand Bank of the coast of Newfoundland and every other offshore fishing area were now off limits.

In effect, the law amounted to a complete moratorium on the colonial New England cod fishing industry. Again, there is no doubt that this was done to discipline the colonists. Once they laid down their weapons, halted their boycotts, and acknowledged Parliamentary sovereignty, there is reason to believe that the British government would have permitted colonists in New England to fish again on a more limited basis in certain areas. The carrot here was North's provision that the Act would only remain in force "till the New Englanders should return to a sense of their duty to the mother country, and submit to her supreme authority." There was reason, however, for colonists to believe that the moratorium would cause purchasers in overseas markets to look for and come to rely on other suppliers. After all, the English Civil War provided just such a window of opportunity for colonial entrepreneurs. Moreover, it soon became clear that there was no chance colonists were ever going to regain legal fishing rights to the Grand Bank, which was the premier fishing ground in the eighteenth-century Atlantic world.

The Royal Navy operating off the coast of North America was ordered to enforce the Act.[33] The navy was "required and enjoined to examine, search, and visit, all ships and vessels carrying on the said fisheries," and to "seize, arrest, and prosecute" those vessels "which shall not have on board the certificate herein before required." The Act mandated that any persons providing customs agents or naval personnel with a "false certificate, cocket, or clearance, for any of the purposes required or directed by this act, such persons shall forfeit the sum of five hundred pounds, and be rendered incapable of serving his Majesty, his heirs, and successors, in any office whatsoever." The loss of £500 was also required for every

[32] *AA*, Series 4, Vol. 1, 1693–1694.

[33] By June 30, 1775, there were twenty-nine British warships stationed off the coast of North America between Florida and Nova Scotia. These warships carried a total of 584 guns and 3,915 men. *NDAR*, Vol. 1, 785. For evidence that these warships enforced North's Restraining Act, see ibid., 1081–1082.

offence of "counterfeiting, erasing, altering, or falsifying, any certificate, cocket, or clearance."[34] By November, all of the thirteen North American colonies were considered in open rebellion, and the Act was extended to include even North American colonists outside of New England.[35] This extension was justified in the same manner as the previous extension to all of New England. It was mandated to defend the principle of Parliamentary sovereignty and for fear that the vicinity of other colonies would enable colonial entrepreneurs to prosper.

The British state then took immediate measures to replace the colonial presence at Newfoundland and to provide state support for the commercial fishing interests in the mother count. In April 1775, Lord North convened a special fisheries committee that resolved to use public funds to subsidize the West Country's migratory cod fishing industry for a period no less than eleven years. This subsidy took the form of bounties. North proposed that the first twenty-five fishing vessels above fifty tons burden to leave from England, or British possessions "in Europe," and return from Newfoundland with at least 10,000 cod caught on the banks, and then make a second fishing expedition, would earn £40. The next 100 vessels to make these two trips would earn £20 each, while the following 100 vessels to do likewise were to earn £10 each. North maintained that "the design" of these bounties was to encourage migratory fishing vessels "going out early enough to make two voyages a year." He explained that "it would be infinitely for the advantage of this country [i.e. England], to make Newfoundland as much as possible an English island, rather than an American colony." He further noted that these bounties would encourage "the bank ship fishery," or the West Country migratory fishing industry, which he personally considered "the great nursery of seamen." The British state allocated £4,000 to encourage West Country fishing interests to catch more fish while the New England commercial fishing industry was closed.[36] These measures were effective, at least initially. A Boston newspaper reported on July 3, 1775, news "from Newfoundland... that a greater Number of Vessels had arrived from Europe for the Fishery than usual, as the Colonies were not permitted to Fish on the Banks there."[37]

[34] *AA*, Series 4, Vol. 1, 1696.

[35] *PDBP*, Vol. 6, 277–278, 290–291.

[36] *PDBP*, Vol. 6, 24; and *AA*, Series 4, Vol. 1, 1807–1808. This bounty system was codified into law. See *An act for the encouragement of the fisheries carried on from Great Britain*... (London: Printed by Charles Eyre and William Strahan, 1775).

[37] Boston Gazette, July 3, 1775. The British state added additional bounties in 1786. *An Act to amend and render more effectual the several laws now in force for encouraging*

Additional government measures in 1775 encouraged the West Country's migratory fishing industry while the colonial industry was restrained. Parliament made it lawful for merchants in Ireland and the Isle of Man to "export directly" to "Newfoundland, or to any Part of America where the Fishery is now or shall hereafter be carried on," provisions, hooks, lines, netting, and other tools of the trade, provided they were "the Product or Manufacture of Great Britain, or the said Isle of Man." "And whereas it has been a Practice of late Years for divers Persons to seduce the Fishermen, Sailors, Artificers, and others, employed in carrying on the Fishery, arriving at Newfoundland, on Board fishing and other Vessels from Great Britain, and the British Dominions in Europe, to go from thence to the Continent of America, to the great Detriment of the Fishery and the Naval Force of this Kingdom," Parliament made it a crime punishable by a fine of £200 to transport fishermen away from the migratory fishing industry "to any Part of the Continent of America." Desertion from labor contracts involved with the migratory fishing industry was made a crime. "Deserters" were subject to loss of wages, imprisonment, and "to be publicly whipped as a Vagrant, and afterwards to be put on Board a Passage Ship, in order to his being conveyed back to the Country whereto he belongs."[38]

Perhaps the most innovative part of Lord North's extensive new maritime program was the repeal of the "Sixth of Anne," which had prohibited the naval impressment of colonial American seamen.[39] The law was explicitly repealed to encourage the migratory fishery. This repeal formally stated:

> And whereas the Said Privilege or Exemption so given by the said Act to Mariners serving on Board Ships or Vessels employed in any of the Seas or Ports of the Continent of America, or residing on Shore there, is prejudicial to the Fisheries carried on by His Majesty's Subjects of Great Britain and Ireland, and others of his Majesty's Dominions in Europe, and has proved an Encouragement to Mariners belonging thereto to desert in Time of War, or at the Appearance of a War, to the British Plantations on the said Continent of America; be it therefore enacted by the Authority aforesaid,

the fisheries carried on at Newfoundland, and parts adjacent, from Great Britain... (London: Printed by C. Eyre and the executors of W. Strahan, 1786). The new bounties were established for a ten-year period. Thus, in total, the British government subsidized the West Country migratory fishing industry for twenty-one years as a direct result of policy decisions in 1775.

38 *An act for the encouragement of the fisheries carried on from Great Britain*... (London: Printed by Charles Eyre and William Strahan, 1775).

39 For more on the "Sixth of Anne," see chapter five, footnote 40.

> That the said Clause, so far as it relates to the exempting of Mariners or other Persons serving, or retained to serve, in any Ship or Vessel in the Seas or Ports of the Continent of America, or other Persons on Shore there, from being impressed, be and the same is hereby repealed.[40]

In all these ways, then, the British state actively attempted to permanently replace the production of the New England cod fisheries in the Atlantic cod trade. Colonists could not simply sit on their hands and hope for a return to the way things had been.

Opposition to the Restraining Act was immediate and fierce, both in England and in the colonies. On the same day the Act passed the House of Lords, sixteen of the twenty-one dissenting peers signed a public declaration opposing the law. This lengthy exposition directly attacked North's mercantilist position and his vicinity argument. It stated "that government which attempts to preserve its authority by destroying the trade of its subjects, and by involving the innocent and guilty in a common ruin [through North's vicinity justification] . . . admits itself wholly incompetent to the end of its institution." The dissenters also followed the London manufacturers and colonial fish merchants in arguing that without the ability to fish, colonial debtors could not reimburse their creditors in England. "Without their fishery," which drew in the most export revenue into New England, they explained, "that is impossible." "Eight hundred thousand pounds of English property," the peers warned, "belonging to London alone, is not to be trifled with." The dissenters objected to the violation of colonial New England charters, which made them "especially entitled to the fishery," and "which have never been declared forfeited." They further maintained that Lord North's belief in "the cowardice of his Majesty's American subjects," or the idea that colonists would simply acquiesce to the Act, did not "have any weight itself." They asserted that "nothing can tend more effectually to defeat the purposes of all our coercive measures than to let the people, against whom they are intended, know, that we think our authority founded in their baseness; that their resistance will give them some credit, even in our own eyes. . . . This is to call for resistance, and to provoke rebellion." The dissenting peers then objected to the fact that "the interdict from fishing and commerce" could only be terminated once royal governors and customs officials confirmed colonial misbehavior had stopped. The peers believed giving governmental "subordinates" the power to "deliver over several hundred thousands

[40] *An act for the encouragement of the fisheries carried on from Great Britain* . . . (London: Printed by Charles Eyre and William Strahan, 1775).

of our fellow-creatures to be starved... without any rule to guide their discretion, with [out] any penalty to deter from an abuse of it," was "a strain of such tyranny, oppression, and absurdity, as we believe never was deliberately entertained by any grave assembly." The peers believed the Restraining Act mandated that the colonies "submit to be the slaves, instead of the subjects, of Great Britain." Furthermore, those signing the public declaration objected to the presentation of the Fisheries Bill as "a measure of retaliation" for colonial disobedience, "upon a supposition that the colonies have been the first aggressors." For the dissenters, colonial non-importation associations and public demonstrations were "no more than a natural consequence of antecedent and repeated injuries." They viewed the Act, therefore, not as "a measure of retaliation," but as "a most cruel enforcement of former oppressions." In their prophetic opinion, the "consequence" of the Act would not be colonial submission, but "a civil war, which may probably end in the total separation of the colonies from the mother country."[41] The peers' protests were published in colonial newspapers.[42]

Individual peers also gave separate fiery speeches in Parliament, protesting the Restraining Act. Lord Camden believed the bill would immediately precipitate war between the colonies and the mother country. After emphasizing the word "war" several times, Camden stated he "knew no other name to give it, for that it carried famine, the worst engine of war, into the bowels of the country [i.e. colonies], and consequently demanded resistance from the tamest and most servile of mankind."[43] Lord Rockingham, a Whig leader during the imperial crisis, believed the Bill to be one of the foremost causes of the American Revolution. The House of Lords debated formal approval of the Tory "An Address of Thanks of the King," in October 1776, and Rockingham proposed replacing the obsequious document with a list of the reasons North's

[41] *PDBP*, Vol. 5, 584–585.

[42] *Providence Gazette*, May 20, 1775; *Pennsylvania Evening Post*, May 11, 1775; and *Protest of the Lords*... (New York: no publisher given, 1775), Early American Imprints, Series I: Evans #14091.

[43] *Lord Camden's Speech on the New-England Fishing Bill* (Newport, RI: S. Southwick, 1775), Early American Imprints, Series I: Evans #42788. Camden's use of the word "famine" here should not be taken as an argument that the Bill would prevent New Englanders from consuming fish. Yankees used dried, salted cod in trade with Pennsylvania for wheat, flour, and bread; and with Virginia and Maryland for grains, pork, and other provisions. New England imported much of its food on the eve of the Revolution. John J. McCusker and Russell R. Menard, *The Economy of British America, 1607–1789* (University of North Carolina Press, 1985), 101, 106, 110.

administration was responsible for bringing about the imperial crisis. In no uncertain terms, Rockingham explained that colonial "seamen and fishermen being indiscriminately prohibited from the peaceable exercise of their occupations, and declared open enemies, must be expected, with a certain assurance, to betake themselves to plunder, and to wreak their revenge on the commerce of Great-Britain."[44]

The Act generated widespread animosity among colonists. Most interpreted the Act as a direct assault on colonial commerce. Leading lights John Adams and John Jay are exemplary on this score. Adams referred to the Act as the "Piratical Act; or Plundering Act," which he held to be primarily responsible for "sundering this country from that [i.e. Britain], I think, forever."[45] Jay believed that the central "question" the Act raised was "whether we shall have trade or not?" He explained in words pregnant with the long-standing belief that fisheries were nurseries of seamen that the Act introduced "a most destructive scheme, a scheme which will drive away all your sailors, and lay up all your ships to rot at the wharves."[46]

The Act specifically threatened to unhinge commercial life in New England, as it directly impacted the region's most important industry. The Act goes a long way toward explaining why the American Revolution began in New England, and why more New Englanders participated in the Revolutionary War than any other regional group. Ultimately, the Act motivated a cross-section of colonial society, entrepreneurs and laborers alike, to fight against the British Empire.

Workers in this region were threatened with the loss of a singular source of employment. The cod fisheries represented jobs to thousands and income to their families. These were jobs in an industry workers had risked life and limb to build. Moreover, colonial workers understood that their economic interests were closely tied to the prospects of the cod fisheries. Alternate forms of employment were few and far between in late-eighteenth-century New England. The amount of available farmland was rapidly shrinking. Farm managers were taking on fewer and fewer workers. Earlier population booms further ensured that there were more

[44] *AA*, Series 5, Vol. 3, 981, 984. The entire text of Rockingham's incendiary "amendment" to "An Address of Thanks of the King" was printed in colonial newspapers. See *The Norwich Packet and the Connecticut, Massachusetts, New-Hampshire, and Rhode-Island Weekly Advertiser*, April 7–14, 1777.

[45] *AA*, Series 4, Vol. 5, 472.

[46] *JCC*, Vol. 3, 495.

workers than jobs in the area.[47] Thus, despite the fact that workers had lost control over the means of production in the industry, and despite the diverging social interests of fishermen and fish merchants in coastal communities, the moratorium on cod fishing surely rang like a fire bell in the night to maritime laborers in New England.

For their part, increasingly wealthy colonial fish merchants had much to lose leaving overseas orders unfilled, waiting for competitors to step in and fill the void the way colonists did during the English Civil War. Colonial entrepreneurs had taken great risks in investing large amounts of capital in the Atlantic cod trade. They had gained access to rich fishing waters and succeeded in out-producing imperial rivals. They even succeeded in gaining control over certain overseas markets. And the entire enterprise had been streamlined and made profitable. The Act placed all of this on the edge of a precipice, and it violated entrepreneurial sensibilities pertaining to economic freedoms.

The loss of the fishing industry also posed a threat to New England businesses linked to the cod fisheries. Shipbuilding, ropewalks, lumber mills, and rum distilleries, hallmarks of New England enterprise, stood to lose much business as a result of the Act. Regional artisans such as carpenters, sailmakers, blockmakers, and blacksmiths also faced reduced demands for their talents.

Rather than quietly accept economic ruin or hope for the best, many colonists in New England mobilized for war. One "Captain Johnson," for example, "a merchant in great repute" in Portsmouth, New Hampshire, cited the Act as the primary reason for his militant resistance to British authority during the Revolution: "The Restraining Act taking place, and depriving him from carrying on his trade, he resolved to go to sea in a privateer, and accordingly hired the *Yankee*, for that purpose."[48]

The Act even helped stimulate revolutionary momentum outside of New England. The Brooklyn town meeting that established New York's Provincial Congress on May 20, 1775, cited the "oppressive" and "evil" prohibition placed on New England's "natural and acquired right of fishing" as key evidence of Parliamentary tyranny. The Act is the only piece of legislation discussed at any length in this important meeting. The

[47] For more on these points, see Winifred Barr Rothenberg, *From Market Places to a Market Economy: The Transformation of Rural Massachusetts, 1750–1850* (University of Chicago Press, 1992); and Vickers, *Farmers and Fishermen*. Rothenberg and Vickers both maintain that the shrinking sources of employment in New England furnished the necessary dependent laborers for late-eighteenth-century industrialization in the region.

[48] *AA*, Series 5, Vol. 1, 755–756.

attendees cited it as justification for forming the colony's Provincial Congress.[49] The New Yorkers explicitly referenced North's vicinity argument, stating their "fear" that they were next to be declared "aiders and abettors of rebellion."[50] Indeed, North's vicinity arguments and his repeated extensions of the Act's jurisdiction meant that they were next.

A committee of safety in North Carolina also took revolutionary action in direct response to the Act. On July 20, 1775, before the extension of the fishing prohibition to all of the thirteen British North American mainland colonies, the committee "Resolved unanimously, That the exception of that Colony, and some others, out of the said act, is a base and mean artifice, to seduce them into a desertion of the common cause of 'America.'" The committee would "not accept of the advantages insidiously thrown out by the said act" in terms of shipping more of its own provisions to the West Indies, and perhaps sending its own vessels to sea to fish. Instead, committee members resolved to "adhere strictly to such plans as have been, and shall be, entered into by the honorable Continental Congress, so as to keep up a perfect unanimity with their sister Colonies." Furthermore, and most provocatively, the committee "unanimously Resolved, not to freight, or in any manner employ, any shipping belonging to Poole, and that they will not carry on any commercial intercourse or communication with the selfish people of that Town."[51] Poole fish merchants, it seems, created many colonial enemies in their efforts to manipulate state power to dominate the Atlantic cod trade.

In sum, the Restraining Act was a mercantilist attempt to kill several birds with one stone. Closing the New England cod fishing industry reduced colonists' ability to purchase and smuggle cheap French sugar. At the same time, the Act reduced West Country merchants' business competition. The Act can also be seen as a disciplinary device aimed at punishing recalcitrant colonists and enforcing Parliamentary sovereignty. Instead of accomplishing these goals, the Act generated widespread animosity both in England and in the colonies.

Griffin and Camden were correct in their concern that the Restraining Act would provoke colonists into resisting British authority through the force

[49] *AA*, Series 4, Vol. 2, 837–838. The New York elites explained natural and acquired rights in the following manner: "natural by their situation, acquired by their joint exertions to acquire the sovereignty of those fisheries." New Yorkers understood and sympathized with fellow colonists' right to use the sea.

[50] Ibid.

[51] *AA*, Series 4, Vol. 2, 1691.

of arms. The Act lit the long fuse of colonial resentment and convinced those involved with the colonial cod fishing industry that the British state no longer supported their maritime interests. To be sure, Parliament did not begin with the goal of destroying the colonial fishing industry. The moratorium imposed on commercial fishing throughout New England was certainly not inevitable. Instead, the Act had roots in a series of state decisions to back West Indian and West Country interests by attempting to command colonists' use of the sea. The subsequent chapters show that colonists mobilized their maritime assets for war rather than submit to the British government's sovereignty of the seas.

PART THREE

THE MILITARY MOBILIZATION OF THE FISHING INDUSTRY

Part Three explores the variety of ways in which maritime assets helped Americans gain their independence.

8

From Trade Routes to Supply Lines

> Sundry Vessels belonging to North America go to the French, Dutch & Danish Islands in these Seas, & offer *unlimited prices* for Gunpowder & other Warlike Stores.... *the vast price offered* may tempt the [West Indian] Proprietors to run risks & dispose of it, which must prove of the utmost disservice to His Majesty by thus assisting the North Americans, who are now declared to be in open Rebellion.[1]

Vice Admiral James Young, commander of the British naval squadron in the West Indies, was very aware that there were vessels bound from the North American mainland colonies entering his jurisdiction to trade illegally for military stores necessary to supply a revolution.[2] Young had been stationed between a rock and a hard place in 1775. On the one hand, commercial vessels owned by law-abiding merchants loyal to Crown and Parliament entered and left his seas on a regular basis. As long as they had the proper papers, this trade was entirely legal, and Young had an obligation to protect and preserve such commerce. On the other hand, New England had been declared in a state of rebellion, and the British government expressly ordered Young to prevent the rebels from obtaining

[1] "Vice Admiral James Young to the President and Members of the Council at Antigua," HMS *Portland*, English Harbor, Antigua, August 14, 1775, *NDAR*, Vol. 1, 1148–1149. Emphasis my own.

[2] The term "military stores" is used here and throughout the chapter to refer to ammunition, chiefly gunpowder, for use in making cartridges and for cannon fodder; flints, for flintlocks; lead, for making musket balls and artillery shot; and arms, including artillery pieces and hand weapons. "Provisions," on the other hand, is used to signify food supplies such as fish, beef, pork, flour, corn, etc. These definitions are in line with late eighteenth-century connotations.

military stores in the West Indies. It was, therefore, left to the naval commander to discern one vessel from the next and to keep arms and ammunition out of New England.

Young's difficult task was made that much more challenging by the fact that the West Indies represented the largest market for dried, salted cod exported from New England in the eighteenth century. Colonial fish merchants had long made profits in trade with merchants and planters in the West Indies. Networks of trust had been established over the decades. Island markets had become dependent on supplies of North American provisions. There were, therefore, numerous New England vessels in West Indian waters.

The commercial ties that bound New England and the West Indies were not severed on Young's watch. Instead, colonial merchants converted traditional overseas trade routes into wartime supply lines for the importation of much needed military stores. Similarly, New England's overseas commercial ties to the Iberian Peninsula were bent to the way of war. Both Atlantic markets played crucial roles in the colonists' conflict with the British Empire.

To fully appreciate how colonial merchants transformed Atlantic shipping lanes into supply lines, and to come to terms with why they were willing to do so, it is necessary to understand that the Massachusetts Provincial Congress controlled much of the military resistance to British authority in 1775. For its part, the Continental Congress convened in Philadelphia first in September 1774 and then again in May 1775.[3] This governing body of Founding Fathers consisted of the educated, wealthy, colonial elite, who were elected to represent the interests of the thirteen North American colonies. Congress created the Continental Army on June 15, 1775. At this time, Congress officially adopted the Massachusetts forces surrounding Boston and chose a Virginian planter and slave owner as commander-in-chief to assuage fears that the war would be fought solely to promote New England's interests. The army was not effectively established, however, until George Washington arrived at Boston on July 2.[4] For the first half of 1775, the Massachusetts Provincial Congress, rather than the Continental Congress,

[3] Edmund Cody Burnett, *The Continental Congress* (New York: MacMillan Company, 1941).

[4] Don Higginbotham, *The War of American Independence: Military Attitudes, Policies, and Practice, 1763–1789*, 2nd ed. (Northeastern University Press, 1983), 85, 98.

was responsible for financing and organizing military resistance to British authority.

Massachusetts's bicameral legislature went through drastic changes between 1774 and 1775. In 1774, councilors appointed by royal writ of mandamus displaced the General Court's upper house to reinforce the royal governor's authority. In response, the popularly elected lower house, the House of Representatives, separated itself and met variously and illegally in Salem, Concord, and Cambridge, under the title "Provincial Congress." In 1775, the Provincial Congress set up permanent shop "at the Meeting-House in Watertown," the epicenter of the Massachusetts forces laying siege to Boston and later General Washington's headquarters.[5]

Following the war's first military engagements in April at Lexington and Concord, alarms were given and armed farmers, and fishermen from the surrounding countryside and coastal areas gathered and bottled up the British forces quartered in Boston. Over the next month, Massachusetts's sons fortified positions around the port city, including Breed's Hill to the north. The creation of these armed units, their military service at the sound of alarms, and their gathering around Boston all occurred under the direction of the Massachusetts Provincial Congress. This was, in effect, a revolutionary government formed by revolutionaries in the midst of military resistance to lead that resistance.

The Massachusetts Provincial Congress established sub-committees to manage the resistance to British authority. A Committee of Safety formed with illustrious members such as John Hancock and James Warren. Committee members oversaw the enforcement of the Continental Association in Massachusetts. In addition, the Committee of Supplies formed to gather military stores and provisions in the event that armed conflict should erupt between the colonists and the British military. The large sum of £20,000 was appropriated specifically for the purchase of arms and ammunition.[6]

Wealthy, politically powerful, fish merchants such as Jeremiah Lee (1721–1775) and Elbridge Gerry (1744–1814) served on the Committee of Supplies.[7] These fish merchants took these important posts at the start

[5] *JHRM*, Vol. 51, Part I, 271; and *JEPCM*, i–iii.

[6] For an overview of the committees of safety and supplies, along with the appropriation figure, see Merrill Jensen, *The Founding of a Nation: A History of the American Revolution, 1763–1776* (Oxford University Press, 1968), 557–560. For the members of the committee of safety in February 1775, see *AA*, Series 4, Vol. 1, 1368.

[7] For evidence that Lee served on this particular committee, see the entry for May 14, 1775, in *JEPCM*, 224; and "Elbridge Gerry, For the Massachusetts Committee of Supplies, to

of military conflict, and they established precedents for the rest of the war. Lee was the son of a wealthy judge and fish merchant. He served as Justice of the Peace, continued his father's business as a fish exporter, and was commissioned Colonel of Marblehead's militia in 1751, a post he would not relinquish until his death. He sat on the Marblehead committee of correspondence; he represented the port town at the Massachusetts Provincial Congress; and he was elected to represent Massachusetts at the Continental Congress. Over the course of his life, Lee owned a total of forty-five ships, brigs, snows, sloops, and schooners. At his death, he left an estate valued at £24,583.18.10.[8] He died suddenly on May 10, 1775.[9]

Gerry, a Harvard graduate, was, like Lee, a proprietor of a fish exporting merchant house inherited from his father. Dr. Benjamin Rush referred to Gerry as "a respectable young merchant, of a liberal education and considerable knowledge. He was slow in his perceptions and in his manner of doing business, and stammering in his speech, but he knew and embraced truth when he saw it."[10] Gerry served as a selectman in Marblehead, a member of the port's committee of correspondence, and a representative to the Massachusetts Provincial Congress. He also signed the Declaration of Independence as a representative to the Continental Congress.[11]

Joseph Gardoqui and Sons, Bilbao, Spain, Watertown, July 5, 1775," in *NDAR*, Vol. 1, 818. For Gerry, see the entry for February 9, 1775, in *JEPCM*, 91; and the "Contract between the Massachusetts Committee of Supplies and Jacob Boardman and others of Newburyport," June 14, 1775, *NDAR*, Vol. 1, 678–679.

8 For biographical accounts of Lee, see Thomas Amory Lee, "The Lee Family of Marblehead," *Essex Institute Historical Collections*, Vol. LII–LIII, (Salem, MA: Newcomb and Gauss, 1916, 1917), 33–48, 145–160, 225–240, 329–344; 65–80, 153–168, 257–287; and *JAB*, II, 665.

9 *JAB*, II, 665.

10 "Sketches of the signers of the Declaration of Independence by Benjamin Rush, c. 1800," in Henry Steele Commager and Richard B. Morris, eds., *The Spirit of Seventy-Six: The Story of the American Revolution As Told by Participants*, 4th ed. (New York: Da Capo Press, 1995), 275.

11 Biographies of Elbridge Gerry, who eventually became Vice President of the United States under James Madison, include George Athan Billias, *Elbridge Gerry: Founding Father and Republican Statesman* (New York: McGraw Hill Book Co., 1976); Samuel E. Morrison, "Elbridge Gerry, Gentleman-Democrat," *NEQ*, II (1929), 6–33; and James T. Austin, *The Life of Elbridge Gerry, With Contemporary Letters, To The Close of the American Revolution* (Boston: Wells and Lilly, 1828). Also, see the short biographical sketches in Clifford K. Shipton, *Sibley's Harvard Graduates* (Harvard University Press, 1970), 239–259; and *JAB*, II, 654. The editors of George Washington's papers maintain that Gerry "was appointed to the committee of supplies on 9 Feb. 1775." *PGW*, Vol. 1,

Fish merchants such as Lee and Gerry used their overseas commercial connections to facilitate the transfer of much needed military stores to the North American colonies. They were well positioned to do so, sitting at the top of the Revolutionary leadership in Massachusetts as members of the Provincial Congress and holding posts on the Committee of Supplies. Given the political power of fish merchants in England and the economic importance of the fisheries, it should not be surprising that fish merchants maintained considerable clout in the colonies.

Gunpowder was one of the most sought-after military stores at the start of the Revolution. Even before the outbreak of war both the colonists and British officials realized that armed resistance of any sort could not succeed without gunpowder. Parliamentarians directly ordered General Thomas Gage to send companies of British regulars into the countryside surrounding Boston to remove caches of colonial powder.[12] At the same time, the British state directed its navy under Vice Admiral Samuel Graves, who was then positioned off the coast of Massachusetts, to stop vessels on the high seas and confiscate any powder thought to be headed into colonial hands.[13] In Gerry's words: "[T]he Ministry in Britain have been endeavoring to keep a Supply of powder from the Colonies, well knowing that they cannot enslave them by any other Means."[14] For their part, the colonists attempted to get their hands on as much powder as they could despite the efforts of the British government and its navy.

In December 1774, the Committee of Supplies began negotiating the purchase of military stores through Spain. Initially, the Spanish monarchy was unwilling to support the colonies openly and thereby risk war with Great Britain. There was, however, no love lost between the two European powers. The British were largely responsible for defeating Spain's ambitions in Florida in 1763. A British settlement in the Falkland Islands seemed to threaten Spanish South America in 1770. Additionally, there was a keen sense of religious animosity between the Protestant

212n. However, according to Gerry, he acted in that capacity from at least as early as December 1774. Compare "Joseph Gardoqui & Sons to Jeremiah Lee, Bilbao, February 15, 1775," *NDAR*, Vol. 1, 401; with "Elbridge Gerry, for the Massachusetts Committee of Supplies, to Joseph Gardoqui and Sons, Bilbao, Spain," Watertown, July 5, 1775, *NDAR*, Vol. 1, 818.

[12] Higginbotham, *The War of American Independence*, 29–53.

[13] Richard Buel, Jr., *In Irons: Britain's Naval Supremacy and the American Revolutionary Economy* (Yale University Press, 1998), 31–32.

[14] "Elbridge Gerry, for the Massachusetts Committee of Supplies, to Joseph Gardoqui and Sons, Bilbao, Spain," Watertown, July 5, 1775, *NDAR*, Vol. 1, 818.

and Catholic empires.[15] For these reasons, the Spanish government did not actively take any steps to prohibit trade with the North American colonies.[16] This tacit support opened the door to entrepreneurial merchants who were willing to take the risks of war-time trade to sell covertly all sorts of goods at inflated prices.[17] Fish merchants on both sides of the Atlantic played leading roles in importing and exporting military stores during the American Revolution.

The most important Spanish merchant house involved in the Atlantic cod trade with New England belonged to the Gardoqui family.[18] Joseph Gardoqui and his sons operated out of Bilbao, Spain. While most Iberian merchants maintained commercial ties to West Country businessmen in England who controlled the Newfoundland fisheries, the Gardoquis developed a transatlantic commercial network, exchanging salt, fruits, wine, and lines of credit for shipments of merchantable grades of dried cod from Massachusetts. Joseph Gardoqui, the patriarch, did business with colonial fish merchants in Marblehead as early as 1741.[19] Bills of exchange were then deposited in accounts with English merchants, mostly in London. Finally, manufactured goods were purchased in England

[15] John H. Elliott, *Empire of the Atlantic World: Britain and Spain in America, 1492–1830* (Yale University Press, 2006); and Jonathan R. Dull, *A Diplomatic History of the American Revolution* (Yale University Press, 1985).

[16] For more on this point, see Nathanael Greene's comments in "Brigadier General Nathanael Greene to Nicholas Cooke," Rhode Island Camp, Jamaica Plains, June 28, 1775, *NDAR*, Vol. 1, 769.

[17] Spanish merchants on the eve of the American Revolution had every reason to hope for lucrative wartime contracts in the event colonial Americans rebelled against Britain. At the beginning of 1775, Elbridge Gerry penned this response to a Spanish merchant's offer of providing military stores in trade: "The Friendship of foreign Factors in this Matter cannot fail of making them respectable & securing to themselves the Interest of these Colonies." The Provincial Congressman further noted that the colonists would "cheerfully allow such a Compensation for your Services as You shall think reasonable." "Elbridge Gerry, for the Massachusetts Committee of Supplies, to Joseph Gardoqui and Sons, Bilbao, Spain," Watertown, July 5, 1775, *NDAR*, Vol. 1, 818.

[18] Reyes Calderón Cuadrado, *Empresarios españoles en el proceso de independencia norteamericana: La casa Gardoqui e hijos de Bilbao* (Madrid: Unión Editorial, 2004). Thomas E. Chávez discusses the Gardoquis only briefly in a more accessible study. Thomas E. Chávez, *Spain and the Independence of the United States: An Intrinsic Gift* (University of New Mexico Press, 2002), 15, 61, 68, 85, 220. Don Diego de Gardoqui represented Spain's interests during peace negotiations toward the end of the Revolution, and he eventually became Spain's first ambassador to the United States. More work needs to be done on the relationship between commercial fishing and the peace process that ended the Revolutionary War.

[19] See "Letter to Joseph Gardoqui," dated Marblehead, May 27, 1741, Joseph Swett Jr. and Robert Hooper Jr. fonds, 1740–1747, MUMHA.

and brought back to the colonies.[20] The Gardoquis also provided colonial merchants with much needed commercial information regarding the prices of various goods and the extent of consumer demand for dried cod in Spain. Indeed, commercial ties between colonial fish merchants and the Gardoqui house were so strong that in 1763 Jeremiah Lee named a ship of 110 tons burden that he used to transport dried cod to Bilbao the *Gardoqui*, in honor of his esteemed trading partners.[21]

The Gardoquis were knowledgeable of colonial grievances against the British government, particularly their frustration with Lord North, and the Spanish merchants were very sympathetic to the colonists' position. On February 15, 1775, months before the outbreak of armed hostilities, the Gardoquis expressed their support in a letter to Jeremiah Lee and the Massachusetts Committee of Supplies. The Gardoquis were "determined at all events to assist you accordingly." They explained:

> We see with the utmost concern the difficulties You labor under by an unpolitical Minister's wrong direction of Affairs, But hope the Present Parliament will look into them with clearer light, & will find proper means to accommodate Matters, without going any further, allowing you your just Rights & Liberties, which we do assure you we long to see it settled with all our hearts; but should it be otherwise (God forbid) command freely and you will find us at your service.... We hourly look out for the *London Post*, should it bring any thing Worth y[ou]r notice, you may depend on being advised.[22]

Two years later, Lord Grantham, the British Minister at Madrid, observed that "the House of Gardoqui is very active. They have long had connection with Great Britain and America, but in the present contest [i.e. the

[20] For examples of these commercial exchanges, see Joseph Swett Jr. and Robert Hooper Jr. fonds, 1740–1747, MUMHA; Richard Derby Ledger, 1757–1776, JDPL; William Knight Account Book, 1767–1781, JDPL; Richard Pedrick Account Book, 1767–1784, MDHS; and Daniel Rogers Account Book, 1770–1790, JDPL. Rogers had accounts with Joseph Gardoqui & Sons that ran from August 1773 to February 1789. Rogers's brig *America* took over shipments of hundreds of quintals of fish, which sold for thousands of pounds. The funds were then deposited in Rogers's account with Champion and Dickason in London.

[21] "A List of Vessels that have cleared outwards in the ports of Salem & Marblehead between the 10th of October 1763 & the 5th of January following being the quarter ending at Christmas..." Massachusetts Shipping Records, 16861765, Part III, 916–1143, box 3, MHS. For more on the cultural significance of ship naming, see Don H. Kennedy, *Ship Names: Origins and Usages During 45 Centuries* (University of Virginia Press, 1974).

[22] "Joseph Gardoqui & Sons to Jeremiah Lee, Bilbao, February 15, 1775," *NDAR*, Vol. 1, 401.

American Revolution], though they pretend to wish it was ended, they have adhered to the latter with great partiality."[23]

The "partiality" that Grantham observed included the fact that the Gardoquis had been shipping military stores to the colonies along with news of British troop movements. On December 16, 1774, four months before the "shot heard around the world," Jeremiah Lee wrote to the Gardoquis to request that the Spanish merchants help the colonists acquire military stores from Europe. Michael Corbet, the Irish mariner who had gained fame for resisting British naval impressment in Marblehead's harbor during the *Pitt Packet* affair in 1769, had been promoted to master of one of Lee's transatlantic vessels. It was Corbet who carried Lee's correspondence across the ocean to the Gardoquis. While Lee's letter has not survived, the Gardoquis responded that they had looked into "the method of complying with your orders . . . to procure as many Muskets & pistols as were ready made." The Spanish merchants had been able to acquire "300 Muskets & Bayonets, & about double the number of Pair of Pistols." They further agreed to do "our utmost to get as many more as was to be found in order to serve you, & shall whenever you should command." Lee and the committee must have also requested gunpowder, as the Gardoquis wrote: "The Powder is an Article which we cannot ship unless we have timely advice, for whatever there is made in this Kingdom is for the Government."[24]

In the summer of 1775, following Lee's death, Gerry and the Committee of Supplies sent "the *Rockingham* [Gerry's vessel, named after the Whig leader in Parliament], Capt[ain] Johnson w[i]th £1000 Ster[lin]g in Cash, & six Hundred & fifty pounds Bills of Exch[ang]e on your House" to the Gardoquis. It is not known whether Gerry was sending his vessel, and presumably his own cash and credit, to fulfill Lee's prior commitment to purchase military stores. Gerry simply wrote that the money was "to be invested in good pistol & Cannon powder – half each."[25] It is known that the *Rockingham* made prior transatlantic trips to Bilbao.[26] It is also

[23] Quote cited in Cuadrado, *Empresarios españoles en el proceso de independencia norteamericana*, 188.

[24] "Joseph Gardoqui & Sons to Jeremiah Lee, Bilbao, February 15, 1775," *NDAR*, Vol. 1, 401. For more on the *Pitt Packet* affair, see the introduction.

[25] "Elbridge Gerry, for the Massachusetts Committee of Supplies, to Joseph Gardoqui and Sons, Bilbao, Spain," Watertown, July 5, 1775, *NDAR*, Vol. 1, 818. The Royal Navy eventually seized the *Rockingham* and confiscated its goods. See Elbridge to Samuel Gerry, May 21, 1776, Elbridge Gerry Papers to 1780, MHS.

[26] Log Book Schooner *Rockingham*, Ship's Logs, box 1, 1926.99. Loan, MDHS.

certain that the Gardoquis continued to play an integral role in supplying military stores to the American fighting forces during the Revolution. John Jay, acting as foreign minister to Spain, wrote to the Continental Congress' Committee of Foreign Affairs from Madrid in 1780 to inform them "the House of Gardoqui at Bilbao are rich, [and] in favor with the Ministry [i.e. the Spanish royal court], and friends to America. The Navy Board have sent to them for goods for the use of the navy, and have remitted to them only an inconsiderable part of the sum to which they will amount, desiring the residue on credit, and promising speedy payment."[27]

The Spanish fish importers also exchanged information pertaining to British troop movements with their transatlantic colonial counterparts. On March 27, 1776, "Joseph Gardoqui and sons" sent a letter by ship to Massachusetts (it took twenty-nine days), where it went through Newburyport, Salem, and finally Watertown, before ending up in the hands of the General Court. The letter stated: "No other news from England, but that 17,300 German troops were going to Boston and Canada, some of which embarking about 3 weeks ago."[28] In this manner, a preexisting transatlantic trade network involving New England and Spanish fish merchants in the exchange of commercial information and merchantable-grade, dried cod was converted to the transfer of arms, ammunition, and military intelligence.

Similarly, prominent colonial fish merchants converted well-worn trade routes to the West Indies into military supply lines.[29] In June 1775, Elbridge Gerry and the other members of the Committee of Supplies signed a contract with "Jacob Boardman Merchant, John Harbert Blockmaker, and Joseph Stanwood Mariner all of Newbury Port in the County of Essex." The contract was to hire "the good Schooner, nam'd the *Britaniae* of the Portlage or Burthen of Eighty-Eight Tons, now laying in the Harbor of said Newbury Port, whereof Abel Coffin is at present Master." The schooner was hired "for a Voyage or Voyages . . . to all or any of the West India Islands, and directly back to any Part of New

[27] Jared Sparks, ed., *The Diplomatic Correspondence of the American Revolution*, Vol. 7 (Boston: Nathan Hale and Gray & Bowen, 1830), 277.

[28] Quote cited in Cuadrado, *Empresarios españoles en el proceso de independencia norteamericana*, 187.

[29] Richard Pares has studied the conversion of trade routes between British North America and the British West Indies into military supply lines for an earlier period. See Richard Pares, *War and Trade in the West Indies, 1739–1763* (Oxford University Press, 1936). To date, there has been no book-length study of a similar conversion during the Revolution.

England" to bring gunpowder to colonial weapons now trained upon Redcoat heads. A provision was added to the contract formally acknowledging it was quite likely "the said Vessel shall be liable or subject to any Seizure, Detention or Confiscation." To mitigate the Newburyport merchants' wartime shipping risks, the committee agreed that the Massachusetts Provincial Congress would be liable for "the Sum of Four Hundred Pounds in full for the Value and worth of the said Schooner, and also the Freight money," the cost of shipping the gunpowder, which amounted to "the Rate of Twenty three Pounds nine shillings and four Pence, for each and every Calendar Month."[30] It is unlikely that these merchants would have conducted this voyage if the revolutionary government did not mitigate risk in this manner.[31] The merchants would not have required such contractual provisions if patriotic devotion to the defense of American liberty had been sufficient motivation in and of itself. They were conditioned to manage the risks associated with maritime commerce in one way or another.

Over the course of the following month, Gerry and the committee arranged for "One hundred & twenty hogsheads Fish, besides Boards, Shingles" to be loaded on board an unnamed brig from Newburyport (not to be mistaken for the schooner *Britaniae*) for a voyage to the Windward islands in the West Indies. Merchants from Marblehead, Gloucester, and Beverly delivered the fish to Newburyport "free of Charge." There, "Captain" Michael Corbet, who had to obtain other employment after Lee's death, was ordered to sell this cargo and obtain as much gunpowder, "an Article, so immediately necessary," as he could among the tropical islands. Gerry and the other members of the committee felt that "the Prosecution of the Scheme in all its Parts, might lay a Foundation for some eminent Services in future."[32] The "Scheme" must have been successful, for in October 1775, the Continental Congress recommended

[30] "Contract between the Massachusetts Committee of Supplies and Jacob Boardman and others of Newburyport, June 14, 1775," NDAR, Vol. 1, 678–679.

[31] For an argument that patriotic, ideologically committed men supplied the American armed forces with food, arms, and ammunition, see E. Wayne Carp, *To Starve the Army at Pleasure: Continental Army Administration and American Political Culture, 1775–1783* (University of North Carolina Press, 1984). Carp focuses on quartermasters in the Continental Army, rather than private contractors.

[32] "Tristam Dalton to Elbridge Gerry," Newburyport, July 23, 1775, *NDAR*, Vol. 1, 953–954. By August, Gerry reported to Washington: "We are in daily Expectation of some Powder from the West Indies, but cannot say what Success our plans will meet with." Elbridge Gerry and the Massachusetts Committee of Supplies to George Washington, Chamber of Supplies, Watertown, August 1, 1775, *PGW*, Vol. 1, 211.

each colony/state export provisions to the West Indies in order to trade for military stores.[33]

As a result of Congress's recommendation, fish merchants were not the only ones to send trade ships to the Caribbean for military stores. Flour and rice merchants in Pennsylvania and South Carolina also had commercial ties to the West Indies, and they too drew upon their connections to gather military supplies.[34] Shippers hid gunpowder in hogsheads topped with sugar, coffee, or some other West Indian trade good.[35] Such techniques increased the likelihood that those vessels the Royal Navy stopped and inspected would evade capture. Throughout much of the war, South Carolina helped supply the New England colonies with imported military stores.[36] Merchants from all over, it seems, recognized the immediate wartime demand for military stores in the New England market. Although the risks associated with overseas trade were heightened during the war as a result of more British warships being on active duty, hope for money overcame the dangers of the sea. Merchants attuned to commercial life in which the greatest risks fetched the most lucrative rewards still sought to meet government contracts with the Massachusetts Provincial Congress and the Continental Congress.

Colonial leaders recognized early on that these profits could be used as a carrot to attract the necessary resources for waging war. At the end of July 1775, the Continental Congress allocated 50,000 dollars to "be paid by the continental treasurers," to merchants in New York and Philadelphia, "to be by them applied to the purpose of importing gunpowder for the continental armies." The exporters were allowed a "five per Cent" commission rate "for their trouble and expenses therein."[37] Congress provided the same five per cent commission to purchasing agents in the West Indies.[38] This gunpowder was to be funneled through various overland and oversea routes to the colonial forces in Massachusetts.

Robert Morris was a wealthy Philadelphia merchant and member of the Continental Congress. His Philadelphia merchant firm received

[33] *JCC*, Vol. 1, 158.

[34] See for example, *NDAR*, Vol. 1, 949, 1013.

[35] *NDAR*, Vol. 1, 929, 949, 1265–1266. J. F. Jameson wrote that Dutch merchants smuggled military stores to the West Indies "disguised in tea-chests, rice barrels, and the like." J. F. Jameson, "St. Eustatius in the American Revolution," *American Historical Review*, Vol. 8, No. 4 (July 1903), 688.

[36] Buel, Jr., *In Irons*, 25–26.

[37] *JCC*, Vol. 2, 210–211.

[38] Robert Greenhalgh Albion and Jennie Barnes Pope, *Sea Lanes in Wartime: The American Experience, 1775–1942* (New York: W.W. Norton and Company, Inc., 1942), 51–52.

Congressional contracts to import arms and ammunition during the Revolution. Morris observed in 1776, "We have a good many imports, but as fast as goods arrive, they are bought up for the Army, or for the use of neighboring States, and therefore continue to bear high prices."[39] War engendered demand for military stores presented opportunities for profits for those with access to government contracts. Thus, the aggressive pursuit of profits that pervaded the eighteenth-century Atlantic world was bent to the way of war in America.

Indeed, the profit motive encouraged merchants to export trade goods, and it prompted purchasing agents to negotiate deals abroad. In July alone, between five and six-and-a-half tons of powder entered Philadelphia from St. Eustatia.[40] At the same time, another 7,000 pounds of powder entered Philadelphia, and 5,000 pounds entered New Jersey.[41] Washington himself believed "there are few Colonies who have not some Vessel out on this Errand [i.e. trading for gunpowder], and will probably bring all that is at Market."[42] In 1776, 69,000 pounds of gunpowder left the West Indies for the mainland colonies.[43] By the end of 1777, 1,454,210 pounds of gunpowder had been imported into the colonies from overseas for the Continental Army.[44]

[39] "Robert Morris to the Commissioners at Paris, Philadelphia, December 21, 1776," Francis Wharton, ed., *The Revolutionary Diplomatic Correspondence of the United States*, Vol. 2 (Washington, DC: Government Printing Office, 1889), 237. For more on Morris's Revolutionary career, see Clarence L. Ver Steeg, *Robert Morris, Revolutionary Financier* (University of Pennsylvania Press, 1954).

[40] For the estimate of five tons, which was provided by an unnamed British colonial official to Philip Stevens, the Secretary of the British Admiralty Board, dated Philadelphia, July 30, 1775, see *NDAR*, Vol. 1, 1013. For the higher estimate, which was recorded in the official journal of the Continental Congress, see *JCC*, Vol. 2 (Washington, DC: U.S. Government Printing Office, 1905), 204. Richard Henry Lee, one of Virginia's delegates to the Continental Congress, wrote to Washington to report the shipment of "six Tons" of powder to Boston. Richard Henry Lee to George Washington, Philadelphia, August 1, 1775, *PGW*, Vol. 1, 209.

[41] See John Adams to James Warren, July 24 and July 27, 1775, *NDAR*, Vol. 1, 964–965, 992; *JCC*, Vol. 2, 238; and Richard Henry Lee to George Washington, Philadelphia, August 1, 1775, *PGW*, Vol. 1, 209.

[42] "George Washington to Nicholas Cooke, Deputy Governor of Rhode Island," Camp at Cambridge, August 14, 1775, *NDAR*, Vol. 1, 1138.

[43] Jameson, "St. Eustatius in the American Revolution," 688. Robert Morris attributed these imports in 1776 in part to the successful "exportation of produce" from New England. "Robert Morris to the Commissioners at Paris, Philadelphia, December 21, 1776," Wharton, ed., *The Revolutionary Diplomatic Correspondence of the United States*, Vol. 2, 237.

[44] O. W. Stephenson, "The Supply of Gunpowder in 1776," *American Historical Review*, Vol. 30, No. 2 (January 1925), 271–281.

Building on the successes of these voyages, a plan was laid for sending a fleet of trade vessels laden with dried cod from Massachusetts to the West Indies to obtain military stores. In December 1775, the Massachusetts Provincial Congress ordered the Committee of Supplies "to engage and fix out with all possible dispatch, on the Account and Risk of this Colony, Ten Vessels, to be laden with as much Provision [including dried, salted cod], Horses, or any other produce of this Colony (except horned Cattle, Sheep, Hogs and Poultry), as they may deem necessary for the importation of Ammunition, Sulfur, Salt-Peter, Arms and German-Steel, for the manufacturing [of] Gun-Locks, &c. And that the said Committee be empowered to give Orders to the Commander of every such Vessel to export the said Cargo to the foreign West-Indies, for the Purposes aforesaid."[45] The Provincial Congress then "put into the Hands of the Committee ... the Sum of Eight Thousand Pounds."[46] Some of this enormous investment went to compensate the committee members for their efforts, just as purchasing agents sent to the West Indies were to be paid for their efforts. This government subsidy provided a financial incentive to colonial fish merchants to utilize their commercial connections around the Atlantic to obtain much needed military stores. Without the fish, the preexisting overseas networks of trust, and the promise of profits, obtaining these stores and prosecuting the war effort would have been much more difficult.

For its part, the British military was aware of the colonial plans. The Royal Navy went to great lengths to stop colonists from importing military stores, particularly gunpowder. On July 5, 1775, British General Thomas Gage, now in command at Boston, communicated to Vice Admiral Samuel Graves, stationed in Boston's harbor, his concern regarding "the Endeavors of the Rebels to procure Powder." Gage informed the Vice Admiral that "within these few Days they have received 37 half Barrels which was brought from New Haven in Connecticut, where it was imported from St. Eustatia."[47] The genial Graves attempted to put a positive spin on matters, responding immediately: "I am happy to find they are in want of Ammunition." In order to explain the situation to the General, the Admiral listed the number of ships under his command and their location. He then wrote: "I anxiously hope to hear that some

[45] *JHRM*, Vol. 51, Part II, 45–46.
[46] *JHRM*, Vol. 51, Part II, 121.
[47] "General Thomas Gage to Vice Admiral Samuel Graves," Boston, July 5, 1775, *NDAR*, Vol. 1, 819.

of them are successful in seizing the Rebel's Supplies, as their Captains are under the strictest Orders to be vigilant, and I omit no occasion of sending Them every information I can procure for their Guidance."[48] For example, on July 5, 1775, Graves ordered Captain William C. Burnaby, commander of the HMS *Merlin*, to cruise "between Cape Anne and Cape Cod [the location of the principal fishing ports in New England], anchoring occasionally at Marblehead or Salem," and also at times "to extend your Cruise to the Isle of Shoals or Piscataqua River." He was especially to "Examine the Bay of Salem and the Harbors therein, in order if possible to prevent every Kind of Supply getting to the Rebels by Sea."[49] The Royal Navy did stop some of the trade in fish for military stores. On August 9, 1775, "two schooners from the West-Indies" attempted to steal into Salem. The HMS *Falcon*, John Lindsay, Captain, captured one of the schooners and chased the second into Salem's harbor, where colonial forces made him withdraw.[50]

For the most part, however, the fish colonists processed over the winter of 1774–1775 was successfully exchanged for military stores at the start of the war. Even Graves had to admit, "On this extensive Coast some Supplies will get in, but I hope in no great Quantities."[51] In the West Indies, on August 30, 1775, Vice Admiral James Young wrote to Philip Stevens, the Secretary of the British Admiralty Board, "that a Schooner from North America arrived at the Island [of] St. Croix, in Order to procure Gunpowder for America; and offering any Money for the Purchase &c., and it having since then been intimated to me by Persons of Credit, at this place, that it is believed The North American Traders do procure Gunpowder and other Warlike Stores from this Island, and His Majesty's other Charibbee Islands, in these Seas, as well as from the French, Dutch, and Danish Islands."[52]

Young ordered the vessels under his command to watch for vessels "bound to North America" laden with "Gunpowder or other Warlike

[48] "Vice Admiral Samuel Graves to General Thomas Gage," Boston, July 5, 1775, *NDAR*, Vol. 1, 819–820.

[49] "Vice Admiral Samuel Graves to Captain William C. Burnaby, R.N.," July 5, 1775, *NDAR*, Vol. 1, 820.

[50] "A Letter from Gloucester, Cape Ann, to Isaiah Thomas, Worcester," Gloucester, August 13, 1775, *NDAR*, Vol. 1, 1132–1133.

[51] "Vice Admiral Samuel Graves to General Thomas Gage," Boston, July 5, 1775, *NDAR*, Vol. 1, 819.

[52] "Vice Admiral James Young to Philip Stevens, Secretary of the British Admiralty," *Portland*, English Harbor, Antigua, August 30, 1775, *NDAR*, Vol. 1, 1267.

Stores."[53] He then wrote to "the President and Members of the Council at Antigua" to warn them that "sundry Vessels belonging to North America go to the French, Dutch & Danish Islands in these Seas, & offer unlimited prices for Gunpowder & other Warlike Stores." Young was "fearful the vast price offered may tempt the Proprietors to run risks & dispose of it, which must prove of the utmost disservice to His Majesty by thus assisting the North Americans, who are now declared to be in open Rebellion."[54] Truly, "the vast price offered" did "tempt" many West Indian merchants with access to military stores to "run risks" to sell military stores, just as it had tempted North American and Spanish merchants. Atlantic businessmen did not stop being entrepreneurs during the war.

American merchants successfully secured military stores during the Revolution for a variety of reasons. Young was the wrong man in the right place. In his own words, he stated, "During the present Hurricane Season I can do little more with the King's Ships under my Command, then sending them to the Southern Islands for Security."[55] Such timid leadership, North America's expansive coastline and numerous harbors, colonial merchants' commercial connections, and "the vast price offered," all contributed to the Patriots' ability to obtain military stores at the beginning of the war.[56]

[53] "Vice Admiral James Young to the Captains of His Majesty's Sloops *Lynx* and *Pomona*," *Portland*, English Harbor, Antigua, August 14, 1775, *NDAR*, Vol. 1, 1148.

[54] "Vice Admiral James Young to the President and Members of the Council at Antigua," *Portland*, English Harbor, Antigua, August 14, 1775, *NDAR*, Vol. 1, 1148–1149.

[55] "Vice Admiral James Young to Philip Stevens, Secretary of the British Admiralty," *Portland*, English Harbor, Antigua, August 30, 1775, *NDAR*, Vol. 1, 1268.

[56] For more on the inability of the British navy to stop military supplies from entering colonial North American ports during the opening salvo of the Revolution, see David Syrett, "Defeat at Sea: The Impact of American Naval Operations upon the British, 1775–1778," in *Maritime Dimensions of the American Revolution* (Washington, DC: Naval History Division, Department of the Navy, 1977), 13–22; Nathan Miller, *Sea of Glory: A Naval History of the American Revolution* (Charleston, SC: The Nautical & Aviation Publishing Company of America, 1974), 22–38; Albion and Pope, *Sea Lanes In Wartime*, 34–64; and Stephenson, "The Supply of Gunpowder in 1776," 271–281. According to Syrett, the British reserved naval power to defend against possible French and Spanish invasion. The British naval leadership derisively discounted the need for naval power in the American theater. The North American coast is enormously long and contains numerous inlets and harbors. Distractions associated with assisting army operations and fending off colonial attacks at sea occupied naval attention. These factors all "made the maintenance of an effective blockade of the American coast impossible." Syrett, "Defeat at Sea," 13–15. More recently, Buel Jr. has demonstrated that the British navy was increasingly successful in the later war years at stopping the flow of goods into and out of the colonies. In particular, he notes that the Royal Navy stationed in the West Indies eventually became proficient at halting French vessels departing from and bound to colonial North American ports. Buel, Jr., *In Irons*, 288, footnote 27.

In sum, American fish merchants converted overseas trade routes into military supply lines during the Revolutionary War. The American leadership successfully harnessed the profit motive of colonial entrepreneurs during the war to secure arms and ammunition from traditional Iberian and West Indian business networks. That is precisely why Vice Admiral James Young was more active in trying to stem the tide of rising prices being offered in the West Indies than he was in stopping colonial shipping on the high seas. He recognized the military and economic imperative of the former, and the futility of the later.

Access to overseas military supplies was crucial to success of the American war effort. Americans simply did not possess the requisite firepower to defeat the British military. If colonists had not been able to make use of seaborne trade routes to import arms and ammunition, then George Washington would have had to order men into battle armed with little more than their own ingenuity and willpower.

9

The First American Navy

> The stagnation of the Fishery furnished us with means for cruising against the enemy's property.[1]

An interventionist British state and the coming of the American Revolutionary War spoiled the lucrative commercial fishing industry in New England. In April 1776, fish merchants in Cape Ann, Massachusetts, estimated that their fleet of "80 fishing Schooners" had declined in value from "£300" to "£150 Each," because they had been idled as a direct result of the war. The merchants claimed a loss of £12,000 for this redundancy, in addition to £6,000 for "Materials for Curing fish; Lost Wholly" and £8,000 "Loss on the Income of 80 Sail fishing Vessels a[t] £100 Each." At the same time, Salem merchants reported that their fleet of fifty "Fishing Vessels" was worth half as much as a result of the war, and they claimed a loss of £7,500 for this depreciation. Salemites further claimed losses of £2,500 for "Materials for curing & Drying fish" that were "Totally Destroyed" and £5,000 for "the Loss of the Fishing 1 Year."[2] For their part, fish merchants in Chatham, Massachusetts, took stock after the war and calculated that they lost twenty-three of their twenty-seven fishing

[1] Resolve of a Boston town meeting at Faneuil Hall, December 11, 1781. *Gentlemen, the inhabitants of the town of Boston...* (Boston: Benjamin Edes and Sons, 1781), Early American Imprints, Series I: Evans #17105.

[2] "Estimate of the Loss on Income & the Trade of Cape Ann from April 1775 to April 1776," and "Estimate of the Loss on Income and Trade for the Town of Salem," April 30, 1776, *NDAR*, Vol. 4, 1323–1325.

vessels as a direct result of the conflict.[3] The large gaps in fish merchants' ledger books between 1775 and 1783 further attest to the war-induced decline in the production of dried, salted cod.[4] Boston merchants further observed in 1781 that "the present decayed state of a number of our maritime towns, contrasted with their flourishing situation previous to the present contest, is a striking proof, not only of the Advantages of the Fishery to the commercial part of the Commonwealth, but the immediate ruin that follows the want of it."[5] In short, the commercial cod fishing industry proved no more resistant to the vicissitudes of war and the British naval blockade than the inland production of colonial grain, flour, and rice.[6]

Yet, this oceanic enterprise did not fade quietly into the night. Not every vessel idled and rotted in harbors. As the Boston merchants who attended the town meeting in 1781 recognized, the disruption of the fishing industry actually provided colonists with the means to achieve some measure of sea power. Fish merchants converted commercial fishing vessels into warships. These hardy vessels constituted an important part of the first American navy.

Colonists armed commercial fishing vessels at the very start of the Revolutionary War in 1775. These warships represent a crucial component of the first American navy for several reasons. First, the mobilization of commercial vessels for war represented part of the American naval strategy developed at the very start of conflict. Second, these vessels were leased directly to the Continental Congress, making them the temporary property of the United Colonies and the first nationally chartered armed

[3] "Description of Chatham, In The County of Barnstable, September, 1802," in *Massachusetts Historical Society Collections*, First Series, Vol. 8 (New York: Johnson's Reprint Corporation, 1968), 153.

[4] See Thomas Pedrick Account Book, 1760–1790, MDHS; Richard Pedrick Account book, 1767–1784, MDHS; Joshua Burnham Papers, 1758–1817, JDPL; William Knight Account Book, 1767–1781, JDPL; Daniel Rogers Account Book, 1770–1790, JDPL; John Stevens Account Book, 1769–75, JDPL; Richard Derby Ledger, 1757–1776, JDPL; and Thomas Davis Account Book, 1771–78, JDPL.

[5] Resolve of a Boston town meeting at Faneuil Hall, December 11, 1781. *Gentlemen, the inhabitants of the town of Boston*... (Boston: Benjamin Edes and Sons, 1781), Early American Imprints, Series I: Evans #17105.

[6] For more on the adverse impact the Revolution had on inland agricultural sectors of the colonial economy, see Brooke Hunter, "Wheat, War, and the American Economy during the Age of Revolution," *WMQ*, Vol. 62, No. 3 (July 2005), 505–526; and Richard Buel Jr., *In Irons: Britain's Naval Supremacy and the American Revolutionary Economy* (Yale University Press, 1998), esp. chapter 1, "The Grain Economy of the Revolution."

vessels. Third, they operated on a basis that defies classification as privateers.

Having said this, the military conversion of fishing vessels was only a part of the process by which American sea power was organized and focused. A navy, after all, is more than a collection of armed ships. The naval strategy that was first developed in 1775, the fishing vessels that were armed for war, the men who manned and commanded those vessels, and the administrative support surrounding them, can *collectively* be seen as the first American Navy.

The process by which an American Navy was formed, much like the formation of an American Army, was such an involved process that it contributed to the establishment of a centralized American government. Food, water, vessels, arms, and ammunition had to be secured and paid for, and these supplies had to be continually restocked.[7] To cover these military costs, the Continental Congress printed money; it borrowed sums from domestic and foreign investors; and it requisitioned funds from the states.[8] The Congress established a bureaucracy to oversee this military spending, and it generated a list of commissioned officers to run the navy. Congress also oversaw the manning of the navy, which entailed recruitment and financing salaries, and it established rules and regulations for American naval warfare.[9] All of this activity helped stimulate the rise of a fiscal and military state in America for the first time in history. This link between naval spending and state formation was not unique to America.[10] Nor was the military mobilization of fishing vessels an innovation, or some part of a distinctly "American" way of war.[11]

[7] Provisioning the navy is discussed in chapter ten.

[8] For more on these financial activities, see E. James Ferguson, *The Power of the Purse: A History of American Public Finance, 1776–1790* (University of North Carolina Press, 1961).

[9] Charles E. Claghorn, *Naval Officers of the American Revolution: A Concise Biographical Dictionary* (Metuchen, NJ: Scarecrow Press, 1988); *Maritime Dimensions of the American Revolution* (Washington, DC: Naval History Division, Department of the Navy, 1977); and Charles Oscar Paullin, *The Navy of the American Revolution: Its Administration, Its Policy, and Its Achievements* (Chicago: The Burrows Brothers, Co., 1906).

[10] See especially Jan Glete, *War and the State in Early Modern Europe: Spain, the Dutch Republic and Sweden As Fiscal-Military States, 1500–1660* (New York: Routledge, 2001); and N.A.M. Rodger, *The Command of the Ocean: A Naval History of Britain, 1649–1815* (New York: Norton, 2005). Also, see John Brewer, *Sinews of Power: War, Money and the English State, 1688–1788* (Harvard University Press, 1990).

[11] For good overviews of the military scholarship devoted to studying the "American" way of war, see "American Military History: A Round Table," *Journal of American*

There is a long Atlantic tradition in which various maritime regions and peoples have used fishing vessels for military/defense purposes. Medieval Vikings used such craft in their frightful raids.[12] In the fifteenth century, indigenous peoples on the West Coast of Africa carved vessels from the trunks of tropical trees and used them along the coastline and inland waterways for a variety of purposes, including fishing. They also mobilized these vessels to ward off well-armed Portuguese invaders.[13] Similarly, North American natives all along the Atlantic seaboard constructed dugout canoes from fallen trees along with kayaks and umiaks from the skins of walrus and seals. These vessels were utilized for a mixture of traveling, trading, fishing, hunting, and fighting off Europeans.[14] The Spanish Armada, which remains one of the most famous naval flotillas in all of recorded history, included Basque, Portuguese, and Spanish fishing vessels.[15] One of the reasons England's Lord Baltimore moved from Newfoundland in the first half of the seventeenth century was the fact that he lost money by having to employ his fishing servants and vessels in war with the French instead of fishing and freighting cod to European markets.[16] England's Rump Parliament relied upon "shallops and ketches," vessels primarily used to catch fish, in addition to ships of the line, to defend its newfound sovereignty from Royalists at home and abroad.[17] Dutch fishing vessels performed military service, and

History, Vol. 93, No. 4 (March 2007), 1116–1162; and Don Higginbothom, "The Early American Way of War: Reconnaissance and Appraisal," *WMQ*, Vol. 44, No. 2 (April 1987), 230–273.

[12] Brian Fagan, *Fish on Friday: Feasting, Fasting and the Discovery of the New World* (New York: Basic Books, 2006), 59–90.

[13] John Thornton, *Africa and Africans in the Making of the Atlantic World, 1400–1800*, 2nd ed. (Cambridge University Press, 1998), 37.

[14] Benjamin W. Labaree, William M. Fowler, Edward W. Sloan, John B. Hattendorf, Jeffrey J. Safford, and Andrew W. German, *America and the Sea: A Maritime History* (Mystic, CT: Mystic Seaport Museum Publications, 1998), 23–28; and Denys Delâge, *Bitter Feast: Amerindians and Europeans in Northeastern North America, 1600–64*, translated by Jane Brierley (University of British Columbia Press, 1993), 51. Also, see Samuel E. Morison, *The European Discovery of America: The Northern Voyages, A.D. 500–1600* (Oxford University Press, 1971), frontispiece, "Frobisher's fight with Eskimos at Frobisher Bay."

[15] Michael Barkham, "Spanish Ships and Shipping," in M.J. Rodríguez-Salgedo, ed., *Armada, 1588–1988* (London: Penguin Books in association with the National Maritime Museum, 1988), 151–163.

[16] Peter E. Pope, *Fish into Wine: The Newfoundland Plantation in the Seventeenth Century* (University of North Carolina Press, 2004), 126.

[17] Bernard Capp, *Cromwell's Navy: The Fleet and the English Revolution, 1648–1660* (Oxford: Clarendon Press, 1989), 4.

many were lost during the first Anglo-Dutch War.[18] Fishing shallops and schooners were also very valuable to the British naval squadron patrolling Newfoundland waters during the 1760s and 1770s for illicit traders, as they enabled naval officers to get in and out of smugglers coves.[19] The Royal Navy likewise found schooners valuable in its efforts to curtail smuggling around Boston on the eve of the Revolutionary War, and some had been previously used as commercial fishing vessels.[20] Virtually all maritime peoples around the Atlantic world relied on fishing fleets for more than catching fish.[21] Thus, the decision to arm fishing vessels during the Revolutionary War was a well-established part of early modern naval conflict.

Colonial Americans did, however, desperately need to establish a naval force to prosecute the Revolutionary War. For the first three years of war, 1775–1777, the American Navy only put four frigates (sizeable warships, with twenty-four to forty-four guns each) to sea, and none of these were operational at the start of the conflict.[22] On the other hand, by June 30, 1775, there were twenty-nine British warships stationed off the North American coast between Florida and Nova Scotia. These warships

[18] Jonathan I. Israel, *The Dutch Republic: Its Rise, Greatness, and Fall, 1477–1806*, 4th ed. (Oxford University Press, 1998), 715–717.

[19] See, for example, NA ADM 1/470/60. For evidence that Commodore Hugh Palliser confiscated French fishing vessels in 1765 and converted them to British warships, see "A List of Vessels & Men, Subjects of France, taken & detained for encroaching on His Majesty's Territories, and for fishing out of the Limits prescribed by Treaties," dated January 1, 1765, prepared by Hugh Palliser, NA SP 78/269/289–291.

[20] The HMS *Gaspée* that was burned on the coast of Rhode Island while pursuing smugglers in Narragansett Bay in 1772 was a schooner. See "Diary of Dr. Ezra Stiles," Newport, May 4, 1775, *NDAR*, Vol. 1, 279. Neil Stout has referred to the schooners patrolling the North American coastline for the Royal Navy as "the original coast guard cutters." Neil R. Stout, *The Royal Navy in America, 1760–1775: A Study of Enforcement of British Colonial Policy in the Era of the American Revolution* (Annapolis, MD: The United States Naval Institute, 1973), 59.

[21] More work needs to be done on the historic relationship between fishing vessels and sea power. In particular, it is not known whether fishing vessels were considered a permanent part of a region's naval power, by its own peoples or its enemies, or if these vessels were merely stop-gap measures pursued by desperate peoples in periods of crisis.

[22] See Richard B. Morris's introduction, *The American Navies of the Revolutionary War* (New York: G.P. Putnam's Sons, 1974), 17. For more on late-eighteenth-century frigates, see William M. Fowler, Jr., *Rebels Under Sail: The American Navy during the Revolution* (New York: Charles Scribner's Sons, 1976), 190–223; and Robert Greenhalgh Albion and Jennie Barnes Pope, *Sea Lanes in Wartime: The American Experience, 1775–1942* (New York: W.W. Norton and Company, Inc., 1942), 22. Fowler places the lower limit of guns typically found on frigates at twenty-four, while Albion and Pope place the upper limit at forty-four. According to Fowler, "only nineteen" frigates actually made it into American naval service over the course of the entire war. Ibid., 190.

carried a total of 584 guns and 3,915 men.[23] Vice Admiral Samuel Graves maintained eight warships under his command for patrolling the New England coastline alone.[24] By October 9, 1775, thirty-five British naval vessels, including twelve ships of the line, patrolled North America's coastline. Fifteen of these warships, and no fewer than seven of the twelve rated warships, were positioned in New England waters at this time.[25] These British vessels served a variety of military purposes. They engaged colonial positions on land, transported and evacuated British troops, and interdicted a portion of the colonial imports and exports. Without some attempt to develop their own sea power, the united colonies could have done little to prevent these things from happening and would have lost the war eventually.

Leading lights in colonial America fiercely debated ways of mitigating Britain's naval advantages, including fitting out vessels for war, from the very start of military conflict in 1775.[26] On the one hand, there

[23] "Vice Admiral Samuel Graves to General Thomas Gage," Boston, June 30, 1775, *NDAR*, Vol. 1, 785.

[24] "Vice Admiral Samuel Graves to General Thomas Gage," Boston, July 5, 1775, *NDAR*, Vol. 1, 819–820.

[25] NA CO 5/122/35.

[26] There is no scholarly consensus on the genesis of the American Navy. Fowler credits Massachusetts with the origins of organized naval resistance to British authority. Fowler, *Rebels Under Sail*, 25–43. Nathan Miller maintains Rhode Island drew up "the first formal movement in behalf of a Continental Navy" on August 26, 1775. Nathan Miller, *Sea of Glory: A Naval History of the American Revolution* (Charleston, SC: The Nautical & Aviation Publishing Company of America, 1974), 41. Credit is also given to Rhode Island in Raymond G. O'Connor, *Origins of the American Navy: Sea Power in the Colonies and the New Nation* (University Press of America, 1994), 15; Kenneth J. Hagan, *This People's Navy: The Making of American Sea Power* (New York: Free Press, 1991), 1; and Frank C. Mevers, "Naval Policy of the Continental Congress," in *Maritime Dimensions of the American Revolution* (Washington, DC: Naval History Division, Department of the Navy, 1977), 3. O'Connor, Fowler, Miller, George Athan Billias, William Bell Clark, Samuel Eliot Morison, and Dudley W. Knox state in no uncertain terms that Washington developed the idea of arming commercial vessels for war. O'Connor, *Origins of the American Navy*, 14; Fowler, *Rebels Under Sail*, 29; Miller, *Sea of Glory*, 60–61; George Athan Billias, *General John Glover and His Marblehead Mariners* (New York: Henry Holt and Company, 1960), 73; William Bell Clark, *George Washington's Navy: Being An Account of his Excellency's Fleet in New England* (Louisiana State University Press, 1960), 3; Samuel Eliot Morison, *John Paul Jones, A Sailor's Biography* (Boston: Little, Brown and Company, 1959), 35; and Dudley W. Knox, *The Naval Genius of George Washington* (Boston: Houghton Mifflin, 1932), 8. Morison goes so far as to refer to Washington as the "'Founder' of the United States Navy." Clark unequivocally states: "General Washington provided the idea." Chester G. Hearn credits Washington and John Glover, a Marblehead fish merchant, with the idea. Chester G. Hearn, *George Washington's Schooners* (Annapolis, MD: Naval Institute

were leaders in the Continental Congress who felt the costs of a naval force outweighed the benefits, while others hoped for reconciliation with the mother country.[27] Samuel Chase, a representative from Maryland, famously stated, "[I]t is the maddest idea in the world to think of building an American fleet.... we should mortgage the whole continent."[28]

Those who supported the formation of an American Navy included Christopher Gadsden, a member of the Continental Congress from South Carolina. Gadsden was the only member of the American government with any formal naval experience. He had been a purser in the Royal Navy and handled ships' money.[29] The South Carolina delegate met John Adams, then acting as Massachusetts's representative at the Congress in Philadelphia. As Adams reported in a letter dated June 7, 1775, Gadsden was "confident that We may get a Fleet of our own, at a cheap Rate." Gadsden believed smaller commercial vessels, such as fishing vessels, could be converted into warships and that the expense of building an entirely new naval fleet could be avoided. Such a "cheap" navy could "easily take their Sloops, schooners and Cutters [smaller vessels], on board of whom are all their best Seamen, and with these We can easily take their large Ships, on board of whom are all their impressed and discontented Men."[30] Gadsden maintained that such pressed men

Press, 1995), 10. For his part, Vincent Dowdell Jr. claims Marblehead merchants developed the idea. Vincent Dowdell, Jr., "The Birth of the American Navy," *United States Naval Institute Proceedings*, Vol. 81 (November 1955), 1251–1257. O'Connor, Mevers, and Miller acknowledge the commander-in-chief's role in fitting out vessels for war at Beverly, Massachusetts, in 1775. However, they lean toward separating Washington's schooners from the American Navy. Mevers's dismissive view is typical on this score: "It is doubtful that Washington intended the squadron to do any more than harass, and it is probably that by this direct action he was demonstrating to Congress his belief in the possibilities of action at sea through a larger maritime force." Mevers, "Naval Policy of the Continental Congress," 3. Mevers credits Congress's Naval Committee and Marine Committee, established between 1775 and 1776, with the formation of the first American Navy.

[27] See the discussion in O'Connor, *Origins of the American Navy*, 16–17.

[28] Quote taken from Hagan, *This People's Navy*, 1.

[29] Stanly Godbold Jr. and Robert Woody, *Christopher Gadsden and the American Revolution* (University of Tennessee Press, 1983).

[30] "John Adams to Elbridge Gerry, Marblehead," Philadelphia, June 7, 1775, *NDAR*, Vol. 1, 628–629. On July 11, 1775, James Warren, President of the Massachusetts Provincial Congress, wrote to Adams to say "I have seen your letter to [Elbridge] Gerry, expressing Mr. [Christopher] Gadsden's opinion about fixing out armed vessels and setting up for a naval power." For his part, Warren wrote that "ten very good [sea] going sloops, from 10 to 16 guns, I am persuaded would clear our coasts. What would 40 such be to the Continent?" *NDAR*, Vol. 1, 857. Adams strongly supported the formation of an American Navy. As early as 1755, he wrote a letter to a friend stating: "All

would not put up much of a fight, especially when pitted against fellow colonists.

Adams then transmitted Gadsden's thoughts to Elbridge Gerry, and the Massachusetts Provincial Congress went back and forth over the issue of arming vessels for war. On June 7, this Congress established a sub-committee of eight merchants "to consider the expediency of establishing a number of small armed Vessels to cruise on our sea-coasts for the protection of our trade and the annoyance of our enemies."[31] In true democratic fashion, Provincial Congressmen debated the issue until shortly after the Battle of Bunker Hill, in which British forces moved troops at will and bombarded colonial positions from the sea. Then, on June 20, 1775, the Provincial Congress resolved "that a number of armed Vessels, not less than six, to mount from eight to fourteen carriage guns, and a proportionable number of swivels, &c be with all possible dispatch provided, fixed, and properly manned, to cruise as the Committee of Safety, or any other person or persons who shall be appointed by this Congress for that purpose, shall from time to time order and direct, for the protection of our trade and sea-coasts against the depredations and piracies of our enemies, and for their annoyance, capture, or destruction." Costs crept back into the discussion, however, and the matter was "ordered to subside *for the present*."[32] This program of arming vessels would resume in Massachusetts later in August.

In the interim, other plans for arming vessels emerged. On June 27, 1775, Nicholas Cooke, the deputy Governor of Rhode Island, wrote to the Massachusetts Committee of Safety to suggest adding "a few Vessels

that part of Creation that lies within our observation is liable to change.... Since we have, I may say, all the naval stores of the nation in our hands, it will be easy to obtain the mastery of the seas, and then the united force of all Europe will not be able to subdue us." Quoted in Richard W. Van Alstyne, *Empire and Independence: The International History of the American Revolution* (New York: John Wiley & Sons, Inc., 1965), 1. Perhaps we might dismiss this as nothing more than the chest-thumping posturing of an ambitious young buck. However, Adams never stopped believing the American colonists could attain "mastery of the seas." See Frederick H. Hayes, "John Adams and American Sea Power," *American Neptune*, Vol. 25, No. 1 (January 1965), 35–45; and Carlos G. Calkins, "The American Navy and the Opinions of One of its Founders, John Adams, 1735–1826," *United States Naval Institute Proceedings*, Vol. 37 (1911), 453–483.

[31] "Journal of the Provincial Congress of Massachusetts," Watertown, June 7, 1775, *NDAR*, Vol. 1, 621–622.

[32] "Journal of the Provincial Congress of Massachusetts," Watertown, June 20, 1775, *NDAR*, Vol. 1, 724. Emphasis my own. O'Connor contends that "this proposal was never implemented." O'Connor, *Origins of the American Navy*, 14. He has not considered the conversion of fishing vessels, however.

properly armed and manned along the Coast in different parts" to protect "our own trade" and to capture "many of the provision Vessels that the Men of War take this way and send round to Boston; many of those Vessels are sent Round [Cape Cod] with but five or six hands in each and With nothing more that a Small arm a-piece to defend them." Cooke further observed that "as the Enemy think we have no force that dare put out of our harbors some of their Store Ships come Without Convoy and there is a possibility that we might pick up one of them if we had a Vessel or two to Cruise in their way."[33] Josiah Quincy wrote to John Adams on July 11, 1775, and suggested "Row Gallies... navigated with many Men... armed with Swivels, and one large battering Cannon in the Bow of each" to "convoy 10 or a Dozen provision Vessels from Harbor to Harbor in the summer Season." He also asked "why might not a Number of Vessels of War be fitted out, & judiciously stationed, so as to intercept & prevent any Supplies going to our Enemies; and consequently, unless they can make an Impression Inland they must leave the Country or starve."[34] All of these plans for arming vessels were sent directly to the Massachusetts Provincial Congress or to John Adams at the Continental Congress, who then forwarded them to the political leaders in his home colony.

The initial colonial naval strategy, then, was worked out between June and July 1775 and transmitted to the seat of war in Massachusetts. The plan at this time involved arming and manning smaller commercial vessels, those under 100 tons, which could be fitted out quickly and at low cost to capture successively larger warships, protect colonial shipping, and cut British military supply lines.

Such a strategy should not be completely reduced to the eighteenth-century way of war known as *guerre de course*, or commerce-raiding, in which enemy merchant vessels were targeted in hit-and-run tactics to bring economic and political pressure to bear on a government through increased maritime insurance rates, price inflation, and shipping losses.[35]

[33] "Nicholas Cooke, Deputy Governor of Rhode Island, to the Massachusetts Committee of Safety," Providence, June 27, 1775, *NDAR*, Vol. 1, 762.

[34] "Josiah Quincy to John Adams, A Massachusetts Delegate to the Continental Congress," Braintree, July 11, 1775, *NDAR*, Vol. 1, 859.

[35] Bernard Brodie, *A Guide to Naval Strategy* (Princeton University Press, 1944), 137. For more on *guerre de course*, see Geoffrey Symcox, *The Crisis of French Sea Power, 1688–1697: From the Guerre d'Escadre to the Guerre de Course* (The Hague: Martinus Nijhoff, 1974). Useful discussions of this naval strategy can also be found in Michael P. Gerace, *Military Power, Conflict and Trade* (London: Frank Cass, 2004), 26–27; Robert Gardiner, ed., *Navies and the American Revolution, 1775–1783* (London: Chatham

The colonists' strategy involved these goals, to be sure. But there were three additional objectives that differentiated the colonial naval strategy from a *guerre de course*. First, colonists hoped to weaken British sea power through the seizure of successively larger warships and the capture of manpower. Second, colonists hoped that cutting British supply lines would cause British forces in Boston to evacuate the port city due to shortages of provisions. Third, colonists believed they could carve a path through the British naval blockade to allow trade to continue unmolested. As a result of these strategic purposes, the fishing fleet that was converted into warships at the end of 1775 must be considered an important component of the first American Navy, rather than mere commerce raiders. These fishing vessels were part of the initial naval strategy worked out in the Continental Congress.

In addition, these fishing vessels were leased directly to the Continental Congress, making them the temporary property of the United Colonies. These leases underscore the vessels' status as the first American warships. On July 18, 1775, the Continental Congress in Philadelphia officially sanctioned the conversion of commercial shipping into armed vessels to meet the aforementioned objectives. The members resolved "that each colony, at their own expense, make such provision by armed vessels or otherwise, as their respective assemblies, conventions, or committees of safety shall judge expedient and suitable to their circumstances and situation for the protection of their harbors and navigation on their sea coasts, against all unlawful invasions, attacks, and depredations, from cutters and ships of war."[36] Marching orders were sent to the Massachusetts

Publishing, in association with the National Maritime Museum, 1996), 66–69; Fowler, *Rebels Under Sail*, 22–23; and Albion and Pope, *Sea Lanes in Wartime*, 25–26. The classic American naval theorist Alfred Thayer Mahan famously argued that a *guerre d'escadre* strategy, which involved targeting an enemy's fighting force, was the only way to achieve command of the sea and victory in war. Alfred Thayer Mahan, *The Influence of Sea Power Upon History, 1660–1783* (New York: Dover Publications, Inc., 1987; originally published by Little, Brown, and Company, Boston, 1890). For his view that commerce raiding did not provide Americans with command of the sea during the Revolution, see Mahan, *The Major Operations of the Navies in the War of American Independence* (London: S. Low, Marston, 1913). Kenneth J. Hagan has recently countered Mahan's argument that Americans only began to realize command of the sea when the nation started to follow a strategy of *guerre de escadre*. For Hagan, a *guerre de course* strategy proved very effective during the Revolutionary War. Hagan, *This People's Navy*, 1–20. While I agree with Hagan that commerce-raiding can be an effective military strategy, we part ways in defining the fishing schooners that were converted into warships during the Revolution as "commerce raiders." Ibid., 2.

[36] *JCC*, Vol. 2, 189.

Provincial Congress, which assigned John Glover, Marblehead fish merchant and colonel of the port's regiment, the task of finding vessels to arm.[37]

In August 1775, Glover succeeded in assembling five of the six vessels the Provincial Congress had resolved back in June to arm. The vessels were all fishing schooners; they all belonged to fish merchants in Marblehead; and they were all converted into warships in Beverly's harbor. The schooners were the *Hannah*, *Franklin*, *Hancock*, *Lee*, and *Warren*.[38]

[37] "Journal of the Provincial Congress of Massachusetts," Watertown, June 20, 1775, *NDAR*, Vol. 1, 724. Glover had been assigned to guard Washington's headquarters and the Provincial Congress at Watertown early in 1775. For more on Glover, see Billias, *General John Glover and His Marblehead Mariners*. Also, see the biographical information compiled in *JAB*, II, 657. At some point he became involved in the search for vessels to convert into warships, although no official document has survived to date Glover's assignment. The Provincial Congress was certainly aware very early on of Glover and his position of authority in the foremost commercial fishing port in New England, and it had relied on him in the past. See, for example, "Minutes of the Massachusetts Committee of Safety," Cambridge, April 27, 1775, *NDAR*, Vol. 1, 229. The Committee, which was affiliated with the Provincial Congress, ordered "That Colonel John Glover" use his authority in Marblehead "for the prevention of Intelligence" leaking to the British patrol vessels in the port's harbor.

On October 4, 1775, Washington assigned Stephen Moylan, the Muster Master General, to assist Glover in arming vessels for war. Both men were to report either to Colonel Joseph Reed, Washington's military secretary, or to the commander-in-chief directly. "Colonel Joseph Reed to Colonel John Glover," Head Quarters, Cambridge, October 4, 1775, *NDAR*, Vol. 2, 289–290; and "Colonel Joseph Reed to Colonel John Glover and Stephen Moylan," Camp at Cambridge, October 4, 1775, in ibid., 290.

[38] The *Franklin*, *Hancock*, *Lee*, and *Warren* were owned by, respectively, Archibald Selman, Thomas Grant, Thomas Stevens, and John Twisden, all Marblehead fish merchants. The *Hancock* was described during the Revolution as "Seventy two Tuns; Taken up for the Service of the united Colonies of America . . . worth Three Hundred Thirty one pounds Six Shillings & Eight pence." "Appraisal of the *Speedwell* [renamed *Hancock*]," Beverly, October 10, 1775, *NDAR*, Vol. 2, 387. At the same time, the *Franklin* was described as "Sixty Tuns; Taken up for the Service of the united Colonies of America . . . worth three Hundred pounds three Shillings and Eight pence." "Appraisal of the *Eliza* [renamed *Franklin*]," Beverly, October 10, 1775, in ibid. The *Lee* was described as "Seventy four Tuns; taken up for the Service of the united Colonies of America . . . worth three Hundred and fifteen pounds Eight Shillings." "Appraisal of the *Two Brothers* [renamed *Lee*]," Beverly, October 12, 1775, in ibid., 412. The *Warren* was described as "Sixty four Tuns; taken up for the Service of the united Colonies in America . . . worth three Hundred & forty pounds ten Shillings." "Appraisal of the *Hawk* [renamed *Warren*]," Beverly, October 12, 1775, in ibid., 412–413. According to the later testimony of a Revolutionary War pensioner, the *Franklin* and the *Hancock* "were Fishing Schooners & had no Bulwarks more than common vessels except Nettings with which they were accustomed to put their clothes in in time of Action." Cited in Philip C. F. Smith and Russell W. Knight, "In Troubled Waters: The Elusive Schooner *Hannah*," *American Neptune*, Vol. 30, No. 2 (April 1970), 27. All except the *Hannah* were re-named, in patriotic fervor, after revolutionary leaders.

Glover leased his schooner *Hannah* of "78 tons" burden to the Continental Congress on August 24.[39] The schooner was built in 1765. Glover purchased it in 1769, and, in typical fashion, the *Hannah* and its crew transported fish and lumber to Barbados in the winter months between 1770 and June 1775, probably having worked the offshore banks in the spring, summer, and fall. The schooner returned bearing muscovado sugar and West Indian rum in her hold.[40] Glover leased the fishing vessel to "the United Colonies of America" or, in other words, the Continental Congress. The Marblehead fish merchant did not lease the schooner to the Massachusetts Provincial Congress, nor did he lease it to General Washington. Such a lease reinforces the *Hannah's* role as the first "American," as opposed to state, naval vessel. And it was not given away freely, which would have indicated that patriotic devotion to the common cause of defending American liberty was the sole motivating factor here. Glover charged the Continental Congress a rate of "one Dollar p[e]r Ton p[e]r Month," or "6" shillings, which, "for two Months & 21 days" amounted to "208 dollars," or £32.8.0.[41] Again, colonial merchants did not stop being entrepreneurs during the Revolution.

It is not known precisely when the *Franklin*, *Hancock*, *Lee*, and *Warren* were armed and officially taken under Washington's command through Glover. Scholars have assumed that the appraisal dates of October 10 and 12 represent the commission dates. See Billias, *General John Glover and His Marblehead Mariners*, 78, 82. However, the communications between Glover and Reed cited above in footnote 27 dispute these October dates. It is probable that the October appraisals were ordered after the vessels had already been secured in Beverly's harbor. Such was the case when the Royal Navy captured the ship *Charming Peggy* on July 15, 1775, and sent her into Boston, where the "Two Thousand one hundred & seventy three Barrels of Flour" could be confiscated for the army. British General Thomas Gage then hired four local merchants to appraise the flour on August 19 to reimburse the flour's owners. The four merchants submitted their appraisal two days later, more than a month after the ship's capture. "General Thomas Gage to Four Boston Merchants," Boston, August 19, 1775, *NDAR*, Vol. 1, 1180.

[39] John Glover's Colony Ledger, MDHS, item 729½. There has been a disagreement about the schooner's size. Fowler describes the *Hannah* as "a typical New England fishing schooner of about seventy tons." Fowler, *Rebels Under Sail*, 29. Hearn, Billias, and Clark follow Glover's Colony Ledger in listing her at "seventy-eight tons." Hearn, *George Washington's Schooners*, 10; Billias, *General John Glover and His Marblehead Mariners*, 74; and Clark, *George Washington's Navy*, 3. Smith and Knight have questioned the use of seventy-eight tons, preferring the much smaller figure of forty-five tons. Smith and Knight, "In Troubled Waters," 15, 22, Appendix II, 41. They base their argument for forty-five tons chiefly on the fact that the terms of Glover's lease add up to around £63, rather than his account of £32.8.0. Here, Glover's words pertaining to his own vessel in his own ledger are taken at face value, and his math skills are discounted.

[40] Smith and Knight, "In Troubled Waters," Appendix II, 41–43.

[41] John Glover's Colony Ledger, MDHS, item 729½. While the amount and the form of payment varied from vessel to vessel, and colony to colony, the rate "per ton per month" was standard. See, for example, "Minutes of the Connecticut Council of Safety,"

Private Joseph Homan, a Marbleheader who described himself in pension records as "a boat fisherman," was one of the maritime laborers responsible for transforming the schooners into warships. Homan testified that in late 1775, "Col. Glover's Regiment was stationed as Marine Corps at the Port of Beverly near Salem for the purpose of manning from time to time small vessels of War, fitted out and manned by the American troops, to intercept and capture British Ordinance vessels and transports bound to the British Army in Boston."[42] Ashley Bowen, a maritime denizen of Marblehead and close observer of town events and town members, recorded in his day book on August 24, 1775, that "a company of volunteers from the Camps at Cambridge" marched through town on their way to Beverly "in order for a cruise in Glover's schooner."[43] Chester G. Hearn has detailed the work done on the *Hannah* at Beverly:

> After tying her up at Glover's Wharf, they cut gunports, two to a side, in her bulwarks and strengthened her planking. For speed, sailmakers increased her usual main, fore, and jib sails by adding topsails and a flying jib. Workmen set a whaleboat amidships and expanded the large cookstove below to serve a larger crew. Glover owned his own cache of arms and provided fours [i.e. four pounders] with carriages, a dozen swivels, and an assortment of gunnery stores.[44]

Glover scrupulously documented the costs he incurred in this conversion process in his ledger-book. On the same day in August that volunteers arrived in Beverly, he charged £151.4.0 to "the United Colonies of America" for provisioning and manning the *Hannah*.[45] Four days later, Glover further billed the Continental Congress £11.9.1 for Ebenezer Foster's blacksmith work on the schooner.[46] Foster mounted swivel guns and did other sundry work on the vessel. In short, Glover did not take a financial loss in converting his fishing vessel into a warship. Because of his lease, Glover stood to gain from transforming his peacetime property into a military machine.[47] Once it had been armed and manned, the

Lebanon, August 3, 1775, *NDAR*, Vol. 1, 1054; and "Stephen Moylan and Colonel John Glover to George Washington," Salem, October 9, 1775, *NDAR*, Vol. 2, 368.

42 Pension Records of the American Revolution, "1818 Pensions," item 1867, the David Library of the American Revolution, Washington Crossing, Pennsylvania.

43 *JAB*, II, 453.

44 Hearn, *George Washington's Schooners*, 10. Also, see Clark, *George Washington's Navy*, 5.

45 John Glover's Colony Ledger, MDHS, item 729½.

46 Ebenezer Foster's blacksmith bill, MDHS, item 5786.

47 The owners of these schooners did not receive prize shares. One-third of the value of the captured vessel and its cargo, whether it was a commercial or a military prize, went to the crew, while two-thirds went to the Continental Congress to repay the cost of outfitting

Hannah set sail for fame and fortune on September 5.[48] Washington took the opportunity to remind Nicholson Broughton, the *Hannah's* captain, in his official sailing orders that it was the American government that paid his salary, not Glover.[49]

In addition to the aforementioned naval strategy and these lease agreements, the fishing vessels operated on a basis that cannot be classified as privateers. William Falconer, the author of a maritime dictionary in 1769, defined a privateer as a privately-owned vessel, fitted out and armed in wartime, "to cruise against and among the enemy, taking, sinking or burning their shipping" in exchange for shares of any captured prizes.[50] To be sure, there is evidence that contemporaries regarded the fleet of armed schooners fitted out at Beverly as a collection of privateers. For example, "Manly, A Favorite New Song in the American Fleet," composed in Salem, Massachusetts, in March 1776, referred to the armed schooner *Lee*, John Manley, Captain, as a "Privateer."[51] Out of exasperation, Washington even went so far as to refer to the men on the schooners as "our rascally privateersmen" in a letter to his secretary Colonel Joseph Reed.[52] Such evidence, combined with the fact that the fishing schooners remained privately owned and the crews (at least) earned prize shares, has led some naval historians to consider these vessels to be mere privateers.[53]

and manning the schooners. "George Washington's Instructions to Captain Nicholson Broughton," September 2, 1775, *NDAR*, Vol. 1, 1288.

48 The *Hannah* is widely touted as the first armed vessel fitted out in the service of the United States. See Hearn, *George Washington's Schooners*, 10; Billias, *General John Glover and His Marblehead Mariners*, 73; Clark, *George Washington's Navy*, 3; *The American Navies of the Revolutionary War*, 22; Smith and Knight, "In Troubled Waters," 29–30; and Knox, *The Naval Genius of George Washington*, 8. Fowler points to earlier "naval actions" off Cape Cod in Buzzards Bay as the genesis of America's naval history. Fowler, *Rebels Under Sail*, 26.

49 "George Washington's Instructions to Captain Nicholson Broughton," September 2, 1775, *NDAR*, Vol. 1, 1287.

50 William Falconer, *A New Universal Dictionary of the Marine* (London, 1769), 353. Gardiner defines privateers as "free-enterprise warships, armed, crewed and paid for by merchants who gambled on the dividend of a valuable capture." Gardiner, ed., *Navies and the American Revolution, 1775–1783*, 66. According to Albion and Pope: "profits were the *raison d'être* of privateers." Albion and Pope, *Sea Lanes in Wartime*, 23–24.

51 *Manly. A favorite new song, in the American fleet. Most humbly addressed to all the jolly tars who are fighting for the rights and liberties of America. By a sailor.* (Salem, MA: Printed and sold by E. Russell, upper end of Main-Street, 1776), Early American Imprints, 1st Series, Evans 43057. Captain Manley's surname may have been deliberately misspelled in the song-title in order to rally men for war.

52 "George Washington to Colonel Joseph Reed," Camp at Cambridge, November 20, 1775, *NDAR*, Vol. 2, 1082.

53 Gardiner references the "handful of Marblehead fishing schooners" in his discussion of "the privateering war." He argues that these schooners do not represent "the beginnings

Following this line of reasoning, the refitted ships were profit-driven, commerce-raiding business ventures, and nothing more.

Having said this, there are several reasons the fishing schooners that were armed for war in late 1775 were not privateers. First and foremost, the vessels themselves were leased to the Continental Congress, which claimed the fleet as its own. Indeed, on November 25, three days before official Naval Regulations were formally approved, Congress specifically referred in its debates over the Regulations to "the Captures heretofore made by Vessels fitted out at the Continental Charge," which at this time pertained to the Beverly fleet.[54] Moreover, the prize money these schooners took went not to the vessel owners, as it would have done with privateers, but rather to the American government to recoup outfitting costs. Also, Washington played a national role in the process as commander-in-chief of the United Colonies by giving out commissions and issuing sailing orders for the schooners. And, in addition to calling the former fishing schooners "rascally privateersmen," the commander-in-chief of the American forces also referred to those same vessels as "the Continental armed vessels" in a letter to the Continental Congress at the end of 1775.[55] Additionally, the crews on the armed schooners were provided wages in addition to prize shares, and these wages were paid by the Continental Congress. The standard practice for privateers in the late-eighteenth century, by contrast, involved giving crews food but not wages.[56] All of this evidence supports the claim that the collection of fishing vessels armed at Beverly represents the first American warships. This fact should not be overly surprising. There was, after all, a well-established naval tradition of arming fishing vessels for war throughout the Atlantic world.

This is not to say that fishing vessels were strictly reserved for what might be considered official naval warfare during the Revolution. Schooners were also widely used as privateers. On November 1, 1775, the Massachusetts Provincial Congress passed an "Act Authorizing Privateers and Creating Courts of Admiralty." The Privateering Act

of a national navy," as "it was conceived with a specific raiding purpose in mind." Gardiner, ed., *Navies and the American Revolution, 1775–1783*, 66–67. For Gardiner, commerce raiding is not associated with official naval activities. For more along these lines, see Hagan, *This People's Navy*, 1–20.

[54] *JCC*, Vol. 3, 376; *NDAR*, Vol. 2, 1133; and L.H. Butterfield, ed., *The Adams Papers: Diary & Autobiography of John Adams, Volume III, Diary 1782–1804, and Autobiography through 1776* (New York: Atheneum, 1964), 349.

[55] General Washington to the President of Congress, Cambridge, December 4, 1775, *AA*, Series 4, Vol. 4, 180.

[56] Albion and Pope, *Sea Lanes in Wartime*, 23.

empowered the Provincial Congress "to Commission, with Letters of Marque and reprisal, any person or persons, within this Colony, who shall at his or their own Expense fix out & equip for the defense of America any Vessel." The Act established a protocol for applying for and receiving letters of marque. This included paying the Provincial Congress's Treasurer a bond of £5,000. The Act also established laws governing the behavior for privateers. It set up admiralty courts to adjudicate the legal seizure of prizes at sea, and the Act codified procedures for selling prize ships and their cargo.[57] Such a system made it possible for fish merchants to apply for letters of marque, convert their fishing schooners, and earn prize money.[58] In contrast to the flotilla armed at Beverly, privateers' owners and crew were allowed to keep 100 percent of their prizes.[59] The owners of privateers also did not pay their crews wages. Such owners were, however, responsible for paying the bond in addition to all of the costs associated with arming and maintaining their vessels.

Merchants' petitions to the Massachusetts Provincial Congress for letters of marque make it possible to discern which fishing vessels were transformed into privateers during the Revolution. Fish merchants

[57] "Massachusetts Act Authorizing Privateers and Creating Courts of Admiralty," November 1, 1775, *NDAR*, Vol. 2, 834–839. Sidney G. Morse has examined the privateer bonds that were filed during the Revolution, and he contends that most bonds were paid not to individual states, but to the Continental Congress. Thus, "Revolutionary privateering" in general, "like the Revolutionary military and regular naval organization, was meant to be a 'Continental' – that is, a *national* – enterprise." Sidney G. Morse, "State or Continental Privateers?" *American Historical Review*, Vol. 52, No. 1 (October 1946), 72.

[58] Albion and Pope define "letters of marque" as "armed merchantmen, licensed by the government to pick up prizes only as by-products of normal trading voyages." They also refer to letters of marque as the official documents licensing these armed merchantmen to take prizes. Albion and Pope, *Sea Lanes in Wartime*, 24–25. However, according to Carl Swanson, most eighteenth-century colonial vessels with letters of marque were privateers. They were not merchantmen primarily concerned with hauling cargoes from one port to another and secondarily able to capture legally the occasional prize. Carl E. Swanson, *Predators and Prizes: American Privateering and Imperial Warfare, 1739–1748* (University of South Carolina Press, 1991), esp. chapter 2 "The Rules of Privateering." The Provincial Congress seems to have broadly interpreted letters of marque to mean licenses for armed commercial vessels to take prizes at any time at sea.

[59] On November 25, 1775, the Continental Congress established formal rules regarding prize shares for privateers, colony/state naval vessels, and Continental naval vessels. The owners of privateers were to get all of the prize money associated with their captures, military or commercial. The colony/state was to get two-thirds of the prize money, and the crew the remainder for their vessels. This same distribution applied to Continental naval vessels, with the Continental Congress getting two-thirds of the prize shares. If, on the other hand, "the Capture be a Vessel of War," then in the case of the colony/state or the Congress, the captors received one-half of the prizes. *NDAR*, Vol. 2, 1133.

in Marblehead, Massachusetts, for example, constantly petitioned the Congress throughout the war for these commissions.[60] In all, the Congress granted nineteen letters of marque to Marblehead merchants. The largest number of these commissions by far was given to commanders of converted fishing schooners. Fourteen of the nineteen commissions, or seventy-four percent of the total, were granted for these vessels between 1776 and 1782.[61] Thirteen schooners belonging to the nearby fishing port of Beverly, Massachusetts, were similarly converted into privateers over the course of the war.[62] In addition, between 1776 and 1783, the Royal Navy captured and sent to the naval base at Halifax, Nova Scotia, 178 colonial vessels. Schooners, the vessel of choice in the cod fisheries, were the most captured vessel type. Eighty-three vessels, forty-seven per cent of the total, were schooners.[63]

A fishing schooner was also used in the first coast guard unit established in Massachusetts in 1775 for the purposes of the Revolution. By August 23, 1775, the people of Machias, in what is today Maine but was then part of Massachusetts's eastern district, had "the armed Schooner *Diligent*" and "the Sloop *Machias Liberty*... fixed for the Purpose of guarding the Sea-Coast." The Provincial Congress put Jeremiah O'Brian in charge of the coast guard vessels and allocated "the Sum of One Hundred and sixty Pounds, Lawful Money of this Colony, for supplying the Men with Provisions and Ammunition." The Provincial Congress also "delivered to the said O'Brian, out of the Colony Store, one Hundred Cannon Balls, of three Pounds Weight each, and two Hundred Swivel Balls; for all which, and the Captures he shall make, he is to account with this Court."[64]

[60] See, for example, a petition dated Boston, November 28, 1777, signed by Samuel Trevett of Marblehead, asking that Captain John Conaway be commissioned as commander of the privateer schooner *Terrible*. *MSSRW*, Vol. 3, 898; and a petition dated Boston, March 29th, 1782, signed by Benjamin Stacey Glover of Marblehead, asking that Benjamin Ashton be commissioned as commander of the privateer schooner *Montgomery*. *MSSRW*, Vol. 1, 317.

[61] Four letters of marque were given to commanders of brigantines, and one vessel was not identified. For the nineteen petitions, see *MSSRW*, Vol. 1, 317; Vol. 2, 448; Vol. 3, 898, 906; Vol. 4, 679, 680, 858; Vol. 5, 277; Vol. 6, 817; Vol. 8, 876; Vol. 9, 588, 624; Vol. 11, 309; Vol. 12, 18, 807; Vol. 13, 982; Vol. 14, 984; and Vol. 16, 109.

[62] This figure was gleaned from the list Howe compiled in *Beverly Privateers in the Revolution*, 405–420.

[63] Albion and Pope, *Sea Lanes in Wartime*, 41.

[64] *JHRM*, Vol. 51, Part I, 96; and *NDAR*, Vol. 1, 1160–1161, 1195. O'Brian made a name for himself as a daring sea captain earlier in May 1775, when he led the force that repelled a British attempt to secure firewood and lumber in Machias. The engagement took place

The same schooner *Diligent* was then used in Massachusetts's own navy. Unlike the schooners fitted out in Beverly that took orders from Washington and were leased to the Continental Congress, and unlike the privateers armed at private expense, the colony/state naval vessels were converted into warships at the colony/state government's expense. The *Diligent* remained the public property of the Provincial Congress, and the warship's crew then took its orders and its pay from this government. Because Washington had appropriated the Beverly schooners, the Provincial Congress formed a special committee on December 29, 1775, "to consider and report" on the possibility of arming additional warships at the Congress's expense.[65] The "Committee for Fitting Out Massachusetts Armed Vessels" reported its findings on January 11, 1776, and the following day the Provincial Congress debated and resolved "that two Ships be built," one of "Thirty-Six Guns" and the other "to carry Thirty-two Guns."[66] However, the Provincial Congress then "re-considered" these resolves on January 29, and at that point "no Establishment has been made for Cruisers."[67] On February 6, the Provincial Congress expanded its naval building program and resolved "That there be built at the Public Expense of this Colony for the Defense of American Liberty, Ten Sloops of War, of One Hundred and Ten Tons, or Fifteen Tons each, suitable to carry from Fourteen to Sixteen Carriage Guns, Six and Four Pounders." Another special committee was then formed "to provide Materials and employ proper Persons to build said Vessels as soon as may be, for the Purpose abovesaid, and that the Sum of Ten thousand Pounds be delivered to the said Committee, to enable them to proceed in building, rigging, and finishing said Vessels as soon as possible."[68]

on land and sea and resulted in the capture of the HMS *Margaretta*, which has been called "the first vessel of the Royal Navy to surrender to an American force." Fowler, *Rebels Under Sail*, 28. Both Fowler and Miller strongly imply that the two Machias vessels and O'Brian were precursors to the American Navy. Fowler, *Rebels Under Sail*, 26–28; and Miller, *Sea of Glory*, 29–35. However, in 1775 these vessels were explicitly armed and manned for the purposes of guarding the coast around Machias.

[65] "Journal of the Massachusetts Council," December 29, 1775, *NDAR*, Vol. 3, 291. The editor, William Bell Clark, considered this date "the beginning of the Massachusetts Navy." Ibid., footnote 2. However, no vessels went to sea at this point. Nor were officers commissioned, nor had pay rates been assigned.

[66] "Report of the Committee for Fitting Out Massachusetts Armed Vessels," January 11, 1776, *NDAR*, Vol. 3, 734.

[67] *JHRM*, Vol. 51, Part II, 219; and *NDAR*, Vol. 3, 1028–1029.

[68] *JHRM*, Vol. 51, Part II, 253–254. This fleet was cut to five sloops on February 16, 1776. See, *NDAR*, Vol. 3, 1315–1316. It was found to be more cost effective to provide leases to vessel owners rather than to build new vessels.

As the fleet expanded, the members of the Provincial Congress decided to use vessels they had previously assigned to guard the coast around Machias as a stop-gap measure. On February 7, 1776, the Provincial Congress resolved that "the Sloop *Machias-Liberty*, and Schooner *Diligent*, now lying at Newbury-Port," be manned with fifty men each.[69] These vessels had been moved from Machias to Newburyport, given new officers and crews, and their mission was formally changed from coast guard duty to colony/state naval service. Another special committee was then established to recommend officers for these vessels.[70] The vessel owners received the same leases as the owners of Washington's schooners, with prize shares going to the crew and the Provincial Congress.[71] This was the beginning of a fleet of vessels that sailed under the Provincial Congress's control.[72]

Converted commercial vessels proved effective weapons of war. From September 1775 to October 1777, the *Hannah* and seven other schooners captured fifty-five vessels, including eighteen brigs, thirteen ships, fourteen sloops, and ten schooners amounting to over 6,500 tons of shipping. In aggregate, British losses amounted to thirty-eight vessels. Eleven captures were judged illegal, because they were found to belong to merchants who were not supplying British forces, and four more were recaptures. The prizes varied in size from a fifteen-ton fishing schooner carrying pilots who were helping the Royal Navy navigate coastal waters, to the 400-ton merchant ship *Concord* carrying dry goods and coal to the British forces

[69] *JHRM*, Vol. 51, Part II, 256; and *NDAR*, Vol. 3, 1156. The idea to use these two vessels in the Massachusetts Navy until the other warships could be built may have come from Jeremiah O'Brian in Machias. See "Petition of Jeremiah O'Brian to the Massachusetts General Court," February 2, 1776, *NDAR*, Vol. 3, 1095. Indeed, O'Brian was ordered to assist the Newburyport Committee of Safety in manning the two vessels. *NDAR*, Vol. 3, 1156; and Vol. 4, 63–64.

[70] *JHRM*, Vol. 51, Part II, 256–257; and *NDAR*, Vol. 3, 1157. The officers were to "be Commissionated by, and follow such Directions as they shall receive from Time to Time from the Council of this Colony."

[71] With regard to its navy, the Massachusetts Provincial Congress did not get around to officially confirming the Continental Congress's late 1775 legislation until February 7, 1776. *JHRM,* Vol. 51, Part II, 256–257.

[72] In 1776, there were three vessels being built in Massachusetts: the brigs *Rising Empire* and *Massachusetts*, and the sloop *Tyrannicide*. *NDAR*, Vol. 4, 259n, 668, 1405; Vol. 5, 1209. Two pre-existing vessels, *Washington* and *Yankee Hero*, were also taken into the Provincial Congress's service at this time. *NDAR*, Vol. 4, 19, 417. "Green & White" uniforms were established for the colony/state navy on April 29, and "the [ship's] Colors be a White Flag with a Green pine Tree, and an inscription 'appeal to Heaven.'" Ibid., 1303.

at Boston.[73] One of the legal prizes, the brigantine *Nancy*, contained a cargo valued at "Fifteen Thousand Pounds Sterling," including:

> 1 Large brass 15 Inch Mortar already fixed for service; a Number of smaller brass mortars; a number [of] Brass Cannon from 24 lb down to 4 lb, with carriages &c all ready; a number of iron cannon of equal size and readiness; 2500 Stand of Arms, Bayonets, &c Pouch's complete &c 30 Tun of [one] ounce [musket] Balls; 10 Tun of swan shot for the muskets; A Great number of carke's already fixed for use to fire the town; Large & small Shot without number; A great number of Hogsheads, filled with cartridges in flannel for the cannon mortars, instead of paper; [and] a great number of Hogsheads of cartridges in paper for the small Arms, with everything of this kind we could have wished for.[74]

These military stores were meant to supply the British forces at Boston. The armed schooners also stopped provisions from reaching the British military in Boston.[75] The timing of these captures was important. Twenty-three of the thirty-eight legal prizes were taken in 1775 at the start of the war. These captured vessels contained military stores, clothing, and provisions worth £31,000, which contributed significantly to the colonists' ability to wage war.[76]

The Massachusetts Provincial Congress also hired fishing vessels for use on special missions. Such was the case with Salem fish merchant Richard Derby Jr.'s schooner *Quero* (most likely named for the fishing

[73] See "Prizes Captured by Washington's Schooners," Hearn, *George Washington's Schooners*, Appendix, 241; and Clark, *George Washington's Navy*, 222–224, and Appendix B, "Prizes Taken by Vessels of Washington's Navy," 229–236. The colonists were also very successful at using whaling boats in coastal raids throughout the siege of Boston. See "Vice Admiral Samuel Graves to Philip Stevens, *Preston*, Boston, July 24, 1775," *NDAR*, Vol. 1, 961–962. More work needs to be done on the whaling industry's military mobilization during the Revolutionary War. Syrett has provocatively referred to "the whaleboat" as "one of the most effective of American weapons systems." Syrett, "Defeat at Sea," 14.

[74] "Edward Green to Joshua Green," Cambridge, December 3, 1775, *NDAR*, Vol. 2, 1247. Brigadier General John Thomas was so excited at this prize that he wrote his wife about it on December 1, 1775: "I have to Inform you of one Thing that is agreeable, viz. one Captain Manly of a Privateer out of Marblehead has brought in to Cape Anne a Fine Large Ship from England Laden with warlike Stores of all Kind Except Powder; a very valuable Prize Indeed; the Particulars I Can't Enumerate. I saw the Invoice which Contains Two Sheets of Paper; this you may Depend that I write nothing of News but what [you] may Rely on for Truth." "Brigadier General John Thomas to His Wife," Roxbury, December 1, 1775, *NDAR*, Vol. 2, 1219.

[75] "George Washington to John Hancock," Cambridge, November 30, 1775, *NDAR*, Vol. 2, 1199. Also, see "Prizes Taken by Vessels of Washington's Navy," Clark, *George Washington's Navy*, Appendix B, 229–236.

[76] Clark, *George Washington's Navy*, 222.

waters known as Banquereau Bank located off Nova Scotia's east coast). Between April and July 1775, the Massachusetts Provincial Congress hired Derby's schooner to take "Depositions relative to [the] Battle of Lexington" to Parliament in London.[77] Similarly, in July, the Provincial Congress approved Salem fish merchant George Dodge's petition "to export 80 Hogsheads of Old Jamaica Cod-Fish from Salem to the West Indies." In return for granting this permission, the Massachusetts political leadership demanded Dodge "order the Master of said Vessel to undertake the Conveyance of a Letter or any other Service that this House may think fit to appoint him to."[78] The Salem merchant captain was thus allowed to set sail for profits in return for carrying the Provincial Congress's correspondence.

The Royal Navy also converted fishing schooners into armed warships during the Revolution. For example, on January 8, 1775, Vice Admiral Graves ordered the purchase of the schooner *Diana* "of 120 tons, about eight Months old, so exceedingly well built that she is allowed to be the best Vessel of the Kind that has been yet in the King's Service; her first cost is £750 Sterling." "I have thought it best for his Majesty's Service that she should be an established armed Schooner," he informed the Secretary of the Admiralty Board; therefore, "I have directed the necessary alterations to be made in her Hull, and for her to be fitted in all respects like other Vessels of her Class; She will have the *St. Lawrence's* Guns." He then appointed one of his family members, Lieutenant Thomas Graves, to command the new warship.[79] By October 1775, five of the thirty-five British warships patrolling the coastline of North America were schooners, most of which had probably been fishing vessels in a former life.[80]

Collectively, the naval strategy that was conceived in 1775, the fishing vessels that were armed for war, the men who manned and commanded those vessels, and the administrative support surrounding them, constitute the first American Navy. The American Revolutionary War transformed the commercial fishing industry in early America in ways that remind one of the manner by which World War II converted American automobile manufacturers into War Department contractors.

[77] "Sch[ooner] *Quero*. Express to England – to Forestall Gen. Gage's Dispatch about the Lexington Fight – (Successful)," *NDAR*, Vol. 1, 967–968.

[78] *JHRM*, Vol. 51, Part I, 15.

[79] "Vice Admiral Samuel Graves to Philip Stevens," *Preston*, Boston, January 8, 1775, *NDAR*, Vol. 1, 59–60.

[80] NA CO 5/122/35.

Whereas twentieth-century automobile magnates received government contracts to produce combat vehicles, the Continental Congress compensated eighteenth-century American fish merchants for transforming fishing vessels into warships. Certainly, these merchants understood that providing Congress with a navy would help the colonies achieve independence, which would ultimately provide them with various political liberties. Yet, these businessmen continued to act as entrepreneurs during the war, just as automobile executives did during the 1940s.

The military conversion of fishing vessels into warships provided a necessary stimulus to the American war effort during the Revolution. Without the military conversion of former fishing vessels, American sea power would have been very limited initially. If fishing vessels had not been armed for war, then the colonists would have been hard pressed to stop the Royal Navy from its military operations in the North Atlantic at the very start of the conflict. Maritime assets were crucial to the American war effort. As the next chapter demonstrates, even fish was mobilized for war.

10

Starving the Enemy and Feeding the Troops

An army marches on its stomach.[1]

Man does not live by bread alone.[2]

While stationed at West Point on a sultry summer day in 1780, American Private Joseph Plumb Martin complained about his rations. Whether his grumblings stemmed from the sun's heat, the food's taste, or the constant complaint typical of soldiers cannot be determined with any precision. Whatever the cause, a disgusted Martin observed, "[O]ur rations, when we got any, consisted of bread and salt shad. This fish, as salt as fire, and dry as bread, without any kind of vegetables, was hard fare in such extreme hot weather as it was then. We were compelled to eat it as it was."[3]

The comments of one common foot soldier underscore a basic fact of war recognized by famous commanders such as Napoleon Bonaparte: troops require food to fight. Shad, a fish commonly found in North American rivers throughout the colonial period, offered sustenance to Martin and enabled him to perform his duties at West Point toward the end of the war, regardless of the fact that he found such food to be "hard fare."

[1] Commonly attributed to Napoleon Bonaparte.

[2] Deuteronomy 8:3.

[3] Joseph Plumb Martin, *Private Yankee Doodle*, ed. George E. Scheer (New York: Acorn Press, 1979), 192. For more on the role that shad played in the Revolutionary War, see Joseph Lee Boyle, "The Valley Forge Fish Story," http://gwpapers.virginia.edu/articles/boyle.html. For more on river fishing in colonial America, see Daniel Vickers, "Those Dammed Shad: Would the River Fisheries of New England Have Survived in the Absence of Industrialization?" *WMQ*, Vol. 61, No. 4 (October 2004), 685–712.

Fish was also brought from the ocean to feed America's fighting men. Dried, salted cod played an important role in provisioning the troops during the Revolutionary War. Like shad, cod was cheaper than meat. The use of both types of fish as a military provision can therefore be seen as a cost-saving measure. There, however, the comparisons end.

Cod was an important Atlantic staple, unlike shad. As a result, there were extra political and economic considerations attached to cod. These additional concerns complicated the use of cod as a provision during the war. On the one hand, colonial leaders did not want fishing equipment, bait, food, rum, and supplies of processed dried cod reaching migratory fishing vessels at Newfoundland. Such a scenario would have had New England fishermen effectively supporting the West Country's Atlantic cod trade. On the other hand, colonial leaders did not want dried, salted cod being traded to Caribbean intermediaries who would then re-export the food to the British military in North America. The American political leadership therefore regulated the use of cod for strategic and economic reasons. In addition, these leaders converted cod, an important oceanic resource, into a fuel source for American fighting men.

Colonial leaders wanted to starve West Country fishermen and British soldiers by depriving them of food and trade goods. Provisions continued to flow out of New England during the Revolutionary War, but these exports were heavily regulated. On May 17, 1775, the Continental Congress resolved to prohibit the direct exportation of "provisions of any kind, or other necessaries" to Newfoundland, Nova Scotia, and other northern territories. This prohibition was explicitly meant to increase the cost of doing business for "the British fisheries on the American coasts."[4]

Loopholes remained in the Restraining Act that created problems with Congress's plan, however. Quaker merchants on Nantucket had received a special exemption for their whaling industry.[5] This meant that

[4] *JEPCM*, 313.

[5] For more on the colonial Nantucket whaling industry, see Lisa Norling, *Captain Ahab Had a Wife: New England Women and the Whalefishery, 1720–1870* (University of North Carolina Press, 2000). As a result of the effective lobby of fellow Quakers in London, Nantucket whalemen were eventually given a special exemption from the general moratorium on fishing. For the exemption, see *AA*, Series 4, Vol. 1, 1694. For evidence of this lobbying, see *PDBP*, Vol. 5, 479–480. Jacob M. Price notes that the economic interests of North American Quakers were exceptionally well represented in Parliament during the eighteenth century. He attributes this representation to the many Quakers who resided in England. Transatlantic Quaker merchants also communicated regularly, and they held a yearly gathering in London. Jacob M. Price, "Who Cared about the Colonies? The Impact of the Thirteen Colonies on British Society and Politics, circa 1714–1775,"

Nantucket whalemen operating off the coast of Newfoundland could sell provisions to English migratory fishing vessels in the region for inflated war time profits, thereby increasing the ability of West Country enterprises to thrive and further expanding the likelihood that the New England fishing industry would be replaced by its inter-imperial commercial rival. Nantucket whalemen were also allowed to enter British-controlled ports in the North Atlantic, further increasing the risk that colonial provisions would end up in the hands of intermediaries who would then supply the British military in Boston. Not everyone was a patriot, and even patriots remained businessmen.

Additional precautions were, therefore, necessary to sew up these loopholes. Continental Congressmen resolved on May 29, 1775, "That no provisions or necessaries of any kind be exported to the island of Nantucket.... The Congress deeming it of great importance to North America, that the British fishery should not be furnished with provisions from this continent through Nantucket, earnestly recommend a vigilant execution of this resolve to all committees."[6] This resolve was then transported to Watertown, Massachusetts, where the colony's Provincial Congress reviewed it on June 9, 1775.[7] Immediately, the Provincial Congress ordered the printing of hand bills to be "dispersed in the several sea port towns in this colony, that due obedience may be paid to the same."[8] These bills stated: "Whereas the Enemies of America are multiplying their Cruelties towards the Inhabitants of the New-England Colonies by seizing Provision Vessels, either the Property of, or intended to supply them... Resolved, That it be, and it hereby is recommended to the Inhabitants of the Towns and Districts in this Colony, that they forthwith exert themselves to prevent the Exportation of Fish, and all other Kinds of Provision."[9] Even before this official prohibition, the Committee of Safety in Marblehead, the foremost commercial fishing port, made it known in town meetings that the Committee would not tolerate supplying the enemy with provisions.[10]

Bernard Bailyn and Philip D. Morgan, eds., *Strangers within the Realm: Cultural Margins of the First British Empire* (University of North Carolina Press, 1991), 423. For more on the Quakers' special ability to get the Empire to work for them, see Alison Gilbert Olson, *Making the Empire Work: London and American Interest Groups, 1690–1790* (Harvard University Press, 1992).

[6] *JEPCM*, 313–314.

[7] *JEPCM*, 313.

[8] *JEPCM*, 314.

[9] *NDAR*, Vol. 1, 608.

[10] Marblehead Town Records, Office of the Town Clerk, Abbot Hall, Marblehead, Massachusetts, January 23, February 14, February 28, 1775.

The Continental Congress did establish an exemption to their prohibition on shipments of provisions to Nantucket. No colony was allowed to supply the Quaker island "except from the colony of the Massachusetts Bay." The Massachusetts Provincial Congress was "desired to take measures for effectually providing the said island [i.e. Nantucket], upon their application to purchase the same, with as much provision as shall be necessary for its internal use, and no more."[11] As the headquarters of colonial whaling operations, Nantucket required food imports. Massachusetts was the only region the Continental Congress allowed to trade with Nantucket because it alone could be trusted. The center of British military operations was in Boston, and Massachusetts was the foremost fishing colony. Massachusetts's merchants were, therefore, the most likely to sell only the provisions that Nantucket needed for local consumption.

Massachusetts's revolutionary government responded by passing a series of resolves to ensure that any provisions shipped to Nantucket were "for the internal use of said island." In addition, the Provincial Congress resolved that no one should be allowed to export provisions to Nantucket "without a permit in writing from the committee of safety of this colony, or such person or persons as they shall appoint to give such permit."[12] Such restrictions were initially considered sufficient.

However, as soon as the people of Nantucket began "fitting out a large fleet of whaling vessels" to exploit the Restraining Act exemption, as they did between June and July, the Provincial Congress felt that it was simply too risky to provide them with provisions. At this point, the Congress resolved on July 8, 1775, "that no provisions or necessaries of any kind be exported from any part of this colony to the island of Nantucket, until the inhabitants of said island shall have given full and sufficient satisfaction to this Congress, or some future house of representatives, that the provisions they have now by them, have not been, and shall not be, expended in foreign, but for domestic consumption."[13] No standards were set for "sufficient satisfaction," however, and even this restriction could not allay fears that provisions would end up in the wrong hands.

Over the course of the next month, Massachusetts's political leaders continued to fret over the Restraining Act's exemption and the shipping of provisions to Nantucket. In the end, the members of the Provincial Congress resolved that only draconian measures would suffice. On August

[11] *JEPCM*, 314.
[12] *JEPCM*, 314.
[13] *JEPCM*, 470.

10, 1775, they decreed "from and after the 15th Day of August Instant, no Ship or Vessel shall sail out of any Port in this Colony, on any Whaling Voyage whatever, without Leave first had and obtained from the Great and General Court of this Colony [i.e. the Provincial Congress]." In effect, Massachusetts did to the Nantucket whalers what Parliament had done to the New England cod fishermen. The Provincial Congress communicated this declaration "to the several Assemblies of the other New England Colonies, advising them to pass a similar Resolve." Then, to make sure the general public was aware of the regulation, the Congress ordered its resolve to "be printed in the several News-Papers of Cambridge and Watertown, and in Hand-Bills." They further established a special subcommittee "for getting the same Printed and Dispersed to the several Sea Ports in this Colony." Members of the Congress justified all of these extreme measures and these attempts to raise public awareness in terms that made it very clear they were trying to starve the Newfoundland fishing interests and the Redcoats in Boston. They stated the need "to take all possible Precaution that none of the Inhabitants of this Colony supply those who are seeking our Ruin, with Provisions, or any Materials that shall enable them to execute their cruel Designs against us."[14] By August 28, New Hampshire's Provincial Congress had also instructed "the Committee of Safety at Portsmouth, to use all prudent methods to hinder any Fish from being exported" without permission.[15] In this manner, the Continental Congress and Provincial Congresses restrained one sector of coastal trade involving shipments of goods and provisions, and they shut down one maritime industry. Americans were asserting control over people's access to the ocean, just as Parliament had done previously.

Surely these actions were taken in the interests of military security. Yet, such measures were also meant to prevent the West Country migratory fishing industry from profiting in the North Atlantic at New England's expense. The commercial competition that existed between the West Country and New England fish merchants before the war continued during the conflict. The regulations placed on cod were, therefore, both strategically and commercially motivated.

Merchants who wanted to trade provisions for military stores in Atlantic waters to the south were more fortunate than those wishing to

[14] *JHRM,* Vol. 51, Part I, 60.
[15] *AA*, Series 4, Vol. 3, 518.

supply the whaling industry. To be sure, there were restrictions on colonial exports in general. On October 20, 1774, the Continental Congress expanded on the original idea of the Society for the Encouragement of Trade and Commerce and established the Continental Association. This non-importation and non-exportation agreement prohibited member colonies from receiving goods from areas Great Britain controlled after December 1, 1774, and it prevented members from shipping provisions to these areas after September 10, 1775.[16]

Fish, however, was still shipped out of New England ports to overseas foreign markets. In addition to the incidents of overseas fish exportation discussed in chapter eight, there is evidence that colonists petitioned for special dispensations to trade dried cod during the war. On June 16, 1775, Ellis Gray and Richard Hinckley, two Massachusetts fish merchants, asked their Provincial Congress for permission to export "eighty hogsheads of Jamaica [i.e. refuse] cod fish, laden on two vessels bound to the West Indies." The petitioners went on to explain that one of the vessels "would have sailed before the resolve of this honorable Congress forbidding the exportation of fish was published, had she not been detained by the elopement of her hands."[17] This shipment of fish was meant to be exchanged for military stores.

Word spread among the merchant community of the availability of foreign military stores in the West Indian islands.[18] It was also widely known that there was burgeoning demand in North America for military stores. Inflated prices for provisions in the West Indies meant that huge war-time profits could be made. For example, the price of dried cod, or "saltfish," increased by eighty-seven per cent per quintal from a peacetime level in 1774 to a wartime level in 1776 in Barbados, one of the largest British markets in the West Indies. In smaller St. Vincent, the price of dried cod increased 300 per cent between 1770 and 1778.[19] Such price

[16] *JCC*, Vol. 1, 75–80; and *NDAR*, Vol. 1, 16n.

[17] *JEPCM*, 343.

[18] Dutch St. Eustatius, in particular, became a major free port at which different European powers could sell military stores to North Americans at inflated prices. J. Franklin Jameson, "St. Eustatius in the American Revolution," *American Historical Review*, Vol. 8, No. 4 (July 1903), 683–708.

[19] Selwyn H. H. Carrington, *The British West Indies During the American Revolution* (Providence, RI: Foris Publications, 1988), 111, table 46b, 113, table 49. Prices for corn, beef, flour, and pork were also inflated throughout the British West Indies. See ibid.; and Richard B. Sheridan, "The Crisis of Slave Subsistence in the British West Indies during and after the American Revolution," *WMQ*, 3rd Series, Vol. 33, No. 4 (October 1976), 615–641, esp. 621, table I. There are no equivalent figures for the French, Spanish,

inflation was a direct result of the Revolutionary War, the Continental Association, and Parliamentary prohibitions. Of course, prices for military stores in the West Indies were inflated due to the exigencies of war, and the inflation of the prices for these goods cut into fish merchants' profit margins.[20] But, the exchange was profitable enough to make wartime trade worth the risks. As a result, then, of colonial demand for military stores and West Indian prices, entrepreneurs such as Gray and Hinckley applied for permission to trade, and they were successful in getting the Massachusetts government to approve their application after certain deliberations were made.[21]

On June 16, the Provincial Congress established a subcommittee "to consider what may be done with respect to such vessels as are now ready to sail, with fish on board, there being a resolve of this Congress against the exportation of fish." This subcommittee on fish exports included two fish merchants from Marblehead, the foremost fishing port: Azor Orne and Jonathan Glover, John Glover's brother.[22] The subcommittee members were tasked with establishing guidelines with which the Provincial Congress could more efficiently evaluate whether future merchants could export fish, earn inflated wartime prices, and bring in much needed military stores despite the prohibitory sanctions. The formation of this subcommittee reinforces the fact that fish exports were highly regulated during the Revolution.

To pave the way for entrepreneurs to export fish in the future, the subcommittee members recommended one particular guideline for granting exemptions for fish exportation during the war. They had examined Gray

or Dutch West Indies. However, it is very likely that some of the same pressures that drove up prices on all the British West Indian islands during the Revolutionary War, particularly the decline in dried cod production, fears of provision shortages, and the increasingly successful efforts of the Royal Navy to halt North American commerce, similarly raised the prices of all imported provisions on the other islands. Reduced provision shipments also inflated prices in Newfoundland. According to West Country fish merchants, declining New England "supplies of bread, flour, and other provisions" to Newfoundland that resulted from the Restraining Act caused "prices of those necessaries of life" to increase "near three times their usual rate." *AA*, Series 4, Vol. 6, 70.

[20] Jameson noted that "it cost in Holland but forty or forty-two florins a hundredweight [for gunpowder], [yet it] brought 240 florins a hundredweight at the island [St. Eustatia]." Jameson, "St. Eustatius in the American Revolution," 688.

[21] The official justification for the permission granted to Gray and Hinckley was that "the fish those gentlemen were about to export, is of such a kind, being old Jamaica fish, as, if detained, cannot possibly be of any advantage to this colony, but must perish." *JEPCM*, 343.

[22] *JEPCM*, 341.

and Hinckley's petition to export refuse-grade cod, and it was their conclusion "that toleration be given to all other owners of vessels, for their departure, who shall convince this Congress, or a committee thereof, that their vessels and cargoes are in the same predicament [i.e. shipping only refuse-grade cod], as no possible advantage can accrue from their detention."[23] The subcommittee's recommendations were formally adopted, and one week later the Provincial Congress approved the petition of Winthrop Sergeant, a fish merchant from Gloucester, Massachusetts, who wanted "to ship off for the West Indies, a quantity of old Jamaica fish, not exceeding forty hogsheads." The language the Provincial Congress used in its approval mirrored the subcommittee's prior recommendation. The Provincial Congress stated that it appeared "to this Congress that the said fish, if stopped, will be of little or no service to this colony."[24] Thus, fish merchants wrote official government policy during the war and made it easier for similar entrepreneurs to export refuse-grade, dried cod, earn profits, and bring in military stores from the West Indies. In this way, cod, an important oceanic resource and Atlantic trade good, was bent to the way of war.

The Massachusetts exemption for allowing the exportation of refuse-grade, dried cod was even linked explicitly to the importation of military stores at the level of the Continental Congress. On July 15, 1775, Ben Franklin, himself a great supporter of Philadelphia's continued trade in bread and flour to the West Indies, secured an exemption allowing all merchants, from any colony, "to load and export the produce of these colonies [later stated "produce of any kind"], to the value of such powder and [military] stores aforesaid, the non-exportation agreement notwithstanding." A copy of this resolution was then printed in the form of hand bills and sent to the West Indies to reassure buyers in the region that they could rely on future North American shipments.[25] This exemption made it acceptable for American merchants to ship provisions to all markets provided they traded for military stores.

Of course, the heads of the British state did not share Franklin's position. As early as 1774 Parliament forbade loyal members of the British Empire to import goods from the rebelling colonies. At the same time, the British state expressly prohibited anyone from exporting military stores

[23] *JEPCM*, 343.
[24] *JEPCM*, 382–383.
[25] *JCC*, Vol. 2, 184–185.

to the colonies without "a License from his Majesty, or the Privy Council for the exportation thereof from Great Britain."[26]

Franklin's exemption, however, effectively removed any legal/political obstacles from the American side. Due largely to his efforts, selling fish to the West Indies was not perceived to violate the spirit of the Continental Association, and this trade was not deemed unpatriotic in the public eye. Thus, dried cod, cattle, grain, and flour from North America continued to flow into the British, French, Dutch, and Spanish West Indies throughout the war for profits and military stores, despite Parliamentary sanctions and British naval patrols.[27]

As a result of greater overseas demand, inflated wartime prices, and Franklin's exemption, the number of petitions to export fish from Massachusetts increased during the war's initial years. Before Franklin's indulgence, there were only six petitions to the Massachusetts Provincial Congress for permission to export fish, five of which were approved.[28] After the exemption, and before France's formal alliance with America in 1778, there were eighteen such appeals.[29]

The Atlantic cod trade thus continued during the war. The trade was circumscribed, involving only those markets willing and able to return military stores. The Revolutionary government further required merchants to seek permission to trade. In addition, for the most part, only

[26] "Narrative of Vice Admiral Samuel Graves," Boston, December 4, 1774, *NDAR*, Vol. 1, 4; and *Boston Evening Post*, December 12, 1774.

[27] Robert Greenhalgh Albion and Jennie Barnes Pope estimate that "four vessels out of five reached their ports safely throughout the hostilities." Robert Greenhalgh Albion and Jennie Barnes Pope, *Sea Lanes in Wartime: The American Experience, 1775–1942* (New York: W.W. Norton and Company, Inc., 1942), 35.

[28] For the other petitions not mentioned above, see *JEPCM*, 377, 413, 419.

[29] Six appeals were granted, three were denied, and the fate of the remainder is uncertain. For the six petitions that were approved, see the petition of Joseph Lee and Nathaniel Tracey, July 26, 1775, in *JHRM,* Vol. 51, Part I, 14, 19; the petition of John Stevens Jr. July 26, 1775, in *JHRM,* Vol. 51, Part I, 14, 18; the petition of Captain George Dodge, sometime in late July 1775, in *JHRM,* Vol. 51, Part I, 15; the petition of Prince Gorham, August 5, 1775, in *JHRM,* Vol. 51, Part I, 46, 47; the petition of "Viellon and Regnault," December 25, 1776, in *JHRM,* Vol. 52, Part II, 199; and the petition of Bartelemi Wasanis, April 3, 1776, in *JHRM,* Vol. 51, Part III, 73–74. For the three petitions whose "Liberty" to export fish was "repealed," see the petition of "Messieurs Pierre la Fitte [Lafitte] and Frederick de la Porte, French gentlemen," April 1, 1777, in *JHRM,* Vol. 52, Part II, 289, and in *JHRM,* Vol. 54, 46; the petition of "Messieurs Domblider and Hurel," September 17, 1778, in *JHRM,* Vol. 54, 43, 46; and the petition of Isaac Smith, April 1, 1777, in *JHRM,* Vol. 54, 44, 46. For the remaining nine petitions, see *JHRM, 1775*, Vol. 51, Part I, 167; *JHRM, 1775–1776*, Vol. 51, Part III, 118, 122; *JHRM,* Vol. 52, Part I, 101; and *JHRM,* Vol. 52, Part II, 115, 199, 286, 288.

refuse-grade, dried cod was allowed to leave the continent in civilian vessels. As a result, the amount of North American provisions entering West Indian waters did not match pre-war levels. As the war dragged on, West Indian planters became increasingly concerned over food shortages and ever more fearful of resulting slave revolts.[30]

In addition to being heavily regulated, fish fed American troops. The conflict's overall cost remained a concern throughout the Revolution, and leaders constantly sought ways to reduce expenditures.[31] Dried, salted codfish helped keep costs associated with prosecuting the war lower than they would have been otherwise.

Fish had been typically less expensive than other primary sources of protein, such as pork and beef, before the war.[32] Saltfish had supplied free and slave workers around the Atlantic World with inexpensive amounts of protein for centuries, and its comparative advantage made it cost effective as a fuel source for fighting men during the Revolutionary War. This was not some form of American ingenuity. European military forces had relied on dried, salted fish for centuries precisely because it was cheap, durable, and portable.[33]

The dried, salted cod used to feed soldiers and seamen during the American Revolution was almost certainly not caught and processed during the military conflict. It is highly unlikely that vessel owners would risk their most capital intensive possessions on fishing expeditions that would have taken months to complete in heavily patrolled waters during a time of war. These entrepreneurs would have further faced the opportunity costs associated with not converting their vessels into warships

[30] For more on the ways the threat to provision imports increased the fears of whites regarding slave revolts, see Andrew Jackson O'Shaughnessy, *An Empire Divided: The American Revolution and the British Caribbean* (University of Pennsylvania Press, 2000), 92–100, 137–143. Fears of food shortages and slave rebellion led Jamaican and Barbadian whites to join with North Americans in protesting the Stamp Act in 1765, and such anxieties even compelled them to protest against the King's decision to wage war against the colonies in 1775. Ibid., 92–96, 141. For more on the Revolution's adverse impact on the West Indian economy, see Carrington, *The British West Indies During the American Revolution*; and Sheridan, "The Crisis of Slave Subsistence in the British West Indies during and after the American Revolution."

[31] For more on these concerns, see E. Wayne Carp, *To Starve the Army at Pleasure: Continental Army Administration and American Political Culture, 1775–1783* (University of North Carolina Press, 1984); and E. James Ferguson, *The Power of the Purse: A History of American Public Finance, 1776–1790* (University of North Carolina Press, 1961).

[32] See the discussion in chapter one.

[33] Brian Fagan, *Fish on Friday: Feasting, Fasting and the Discovery of the New World* (New York: Basic Books, 2006), esp. 123 and 225.

for naval service or privateering. Moreover, workers would have had to have been willing to risk impressment into the Royal Navy for working these post-Restraining Act fares. Fish merchants would have had to offer inflated earnings to these workers as incentives, which in turn would have increased the regular costs of doing business. Then, there is the fact that many fishermen fought in the war, which suggests that most maritime laborers stopped working spring, summer, and fall fares, as these were the seasons for fighting.[34] War-year gaps in fish merchant account books and fishing communities' petitions for poor relief after the war further reinforce the fact that most American commercial fishing expeditions stopped as a direct result of the Restraining Act and the American Revolutionary War.[35]

It is most likely that a majority of the dried, salted cod used to provision the American armed forces during the Revolution had been caught before the summer of 1775. Once processed, spring fare cod, along with left-over fall fare cod from 1774 that had not been shipped overseas, could have endured long periods of storage in warehouses in fishing communities all along the New England coast. One of the unique properties of cod is that its flesh can withstand heavy salting. Such a salt cure can preserve properly stored cod for up to five years.[36]

Marblehead fish merchant William Knight regularly sent shipments of dried, salted cod to the Revolutionary leadership in colonial Massachusetts. During the summer months in 1775, two of Knight's fishing vessels, the schooner *Molly* and the *Barnett*, and two fishing companies working for Knight, transported "spring fare" cod caught before the Restraining Act to "Water Town," the seat of Massachusetts' revolutionary government and General Washington's headquarters. Knight scrupulously recorded the "expenses" he incurred for these shipments.[37] This fish was undoubtedly used to provision the troops then laying siege to Boston. The Provincial Congress' Committee of Supplies had resolved as early as February 22, 1775, to purchase "four hundred and fifty quintals [50,400 pounds] of Saltfish."[38]

[34] The military mobilization of fishermen is discussed in chapter eleven.

[35] For these gaps and petitions, see the introduction to chapter nine.

[36] See the discussion in chapter one.

[37] William Knight Account Book, 1767–1781, JDPL.

[38] *AA*, Series 4, Vol. 1, 1368. By comparison, the British Army's "Provisions in store, at Boston, the 16th November, 1775" included a total of "23,469 lbs. dried cod-fish," roughly 210 quintals. *AA*, Series 4, Vol. 4, 160.

The Continental Army that was eventually formed had more mouths to feed than even the largest urban centers in British North America between 1775 and 1783. Philadelphia was the largest city, with a population of 33,290 in 1775. New York City maintained a population of 21,863 in 1771. Boston supported around 16,000 people in 1775. Throughout the war, the Army's total manpower on a yearly basis was 32,625 in 1775; 78,425 in 1776; 51,175 in 1777; 39,225 in 1778; 36,700 in 1779; 33,825 in 1780; 25,600 in 1781; and 15,250 in 1782. On average, there were 39,103 men in the Army. Thus, there were typically more people in the Army than there were in the leading city of Philadelphia or the entire colony/state of Georgia.[39]

These fighting men required enormous amounts of food. In 1778, the commissary general for the Army estimated that a force of 81,000 would consume 203,000 barrels of flour each year.[40] The demand for this flour contributed in part to the Continental Congress's decision in 1777 to form a permanent army, enlist smaller numbers of men for the duration of the war, and thereby preserve the agricultural workforce for sowing and reaping.

The fighting men did not live on bread alone, however. At the start of the war in June 1775, before the formation of the Continental Army, the Massachusetts Committee of Supplies set down "the allowance for provisions for the soldiers of the Massachusetts army." "Each soldier" was to receive the following daily rations:

> One pound of bread. Half a pound of beef and half a pound of pork; and if pork cannot be had, one pound and a quarter of beef; and one day in seven they shall have one pound and one quarter of saltfish, instead of one day's allowance of meat. One pint of milk, or, if milk cannot be had, one gill of

[39] For Philadelphia, see Billy G. Smith, *The "Lower Sort": Philadelphia's Laboring People, 1750–1800* (Cornell University Press, 1990), table B.1, 206. For New York City, see Edward Countryman, *A People in Revolution: The American Revolution and Political Society in New York, 1760–1790* (Johns Hopkins University Press, 1981), 331, footnote 52. For Boston, see Gary B. Nash, *The Urban Crucible: Social Change, Political Consciousness, and the Origins of the American Revolution* (Harvard University Press, 1979), figure 1, 409; and Carl Bridenbaugh, *Cities in Revolt: Urban Life in America, 1743–1776* (New York: Alfred A. Knopf, 1955), 5. For the Army, see Richard Buel Jr. *In Irons: Britain's Naval Supremacy and the American Revolutionary Economy* (Yale University Press, 1998), table 2.1, 49. The Army figures include all of the men raised for service "from North Carolina to present-day Maine." For its part, Georgia maintained a population of 15,000 in 1775. Robert V. Wells, *The Population of the British Colonies in America before 1776: A Survey of Census Data* (Princeton University Press, 1975), table VII-5, 284.

[40] Buel Jr., *In Irons*, 27.

rice. One quart of good spruce or malt beer. One gill of peas or beans, or other sauce equivalent. Six ounces of good butter per week."[41]

At the same time, the daily rations for troops from Connecticut included the following: three-quarters of a pound of pork or one pound of beef; one pound of bread or flour; three pints of beer "or Spruce Sufficient"; and one pint of milk. Dried, salted cod, or "saltfish," was to be substituted for the daily meat ration three times a week, and each man was to receive half a pint of rice or one pint of meal; six ounces of butter; and three pints of peas or beans each week.[42] "The Ration of Provisions allowed by the Continental Congress unto each Soldier" for the Continental Army included the following: either one pound of fresh beef or three-quarters of a pound of pork or one pound of saltfish per day; one pound of bread or flour per day; three pints of peas, or beans, or "Vegetables equivalent" per week; one pint of milk per day "when to be had"; half a pint of rice or one pint of "Indian meal" per week; and one quart of "Spruce Beer" or nine gallons of molasses "per Company of 100 Men," per day, most likely for small beer. On Monday and Wednesday mornings, soldiers were to get a "ration" of salt and fresh meat and two rations of bread. Every Friday was deemed a fish day, and soldiers were to get a ration of "saltfish."[43] Clearly, none of these soldiers were meant to subsist solely on dried cod. But Massachusetts, Connecticut, and the Continental Congress did recognize the added value in supplementing soldiers' diets with inexpensive, protein-rich fish.

For its part, the Continental Navy did not have nearly as many mouths to feed as the Army. Throughout the war, the Navy's manpower per year was 550 in 1775; 1,910 in 1776; 4,023 in 1777; 2,840 in 1778; 2,428 in 1779; 1,726 in 1780; 1,724 in 1781; and 640 in 1782.[44] On average, there were 1,980 mouths to feed in the navy in any given year, and fish played an important role in feeding them. On October 30, 1775, the Continental Congress appointed seven men to a Naval Committee. Four of the committee members were New Englanders, including John Adams, who was familiar with the commercial fishing industry because he lived in the port community of Quincy, Massachusetts.[45] Adams also represented

[41] *JEPCM*, 317–318.

[42] *PGW*, Vol. 1, 233, footnote 22.

[43] General Orders, Head Quarters, Cambridge, August 8, 1775, *PGW*, Vol. 1, 269; and *JCC*, Vol. 3, 322.

[44] Buel, Jr., *In Irons*, 293, footnote 60.

[45] Raymond G. O'Connor, *Origins of the American Navy: Sea Power in the Colonies and the New Nation* (University Press of America, 1994), 15; and Samuel Eliot Morison,

fish merchants in legal cases and became knowledgeable about the industry to defend his clients better.[46] In his proud words: "My Practice as a Barrister in the Counties of Essex, Plymouth and Barnstable had introduced me to more Knowledge both of the Cod and whale fisheries and of their importance both to the commerce and Naval Power of this Country than any other Man possessed."[47] Hyperbole aside, Adams's intimate knowledge and appreciation of the fishing industry motivated him to insert the following clause in the very first "Rules for the Regulation of the Navy of the United Colonies" on November 28, 1775:

> All ships furnished with fishing tackle, being in such places where fish is to be had, the Captain is to employ some of the company in fishing; the fish to be distributed daily to such persons as are sick or upon recovery, provided the surgeon recommend it, and the surplus, by turns amongst the messes of the officers and seamen, without favor or partiality and gratis, without any deduction of their allowance of provisions on that account.[48]

In addition to moral prohibitions forbidding naval seamen "to swear, curse, or blaspheme the name of God," and formal recognition of "the cat of nine tails" as a disciplinary tool, the fishing clause and other regulations applied to "Commanders of all ships and vessels belonging to the thirteen United Colonies," which retroactively included the armed schooners fitted out at Beverly and leased to the Continental Congress. Those serving on board former fishing vessels, and anyone else the Naval Regulations applied to, were encouraged to catch fish to supplement their diets.

Fish was also used to feed coast guard units. On June 28, 1775, the Massachusetts Provincial Congress resolved to establish permanent "companies on the sea coast." Thirty-five companies, consisting of fifty men each, were raised and stationed in port communities from Plymouth County in southern Massachusetts to Cumberland County in what is

John Paul Jones, A Sailor's Biography (Boston: Little, Brown and Company, 1959), 36.

[46] See, for example, L. H. Butterfield, ed., *The Adams Papers: Diary & Autobiography of John Adams, Volume I, Diary 1755–1770* (New York: Atheneum, 1964), 154–157.

[47] John Adams Autobiography, Part 2, "Travels, and Negotiations," 1777–1778, Sheet 1 of 37, in the Adams Family Papers, MHS.

[48] *JCC*, Vol. 3, 380; and *NDAR*, Vol. 2, 1176, 1179. In addition to fish, naval seamen received daily rations of rum, bread, beef, pork, peas, rice, butter, cheese, turnips, potatoes, and onions. For Adams's formative role in writing the Naval Regulations, see Frederick H. Hayes, "John Adams and American Sea Power," *American Neptune*, Vol. 25 (1965), 36; and Butterfield, ed., *The Adams Papers*, Vol. III, 346, footnote 2.

today Maine.[49] Then, between December 28, 1775, and January 4, 1776, the Provincial Congress debated and resolved to expand the number of "Forces to guard the Sea Coast." A total of 2,650 men were recruited to guard coastal communities between Martha's Vineyard and Falmouth (now Portland, Maine) for the period of one year.[50] Although the port communities themselves were responsible for recruiting the men, the Provincial Congress established the organizational structure for these units. This Congress issued orders for these men, established commissary agents in certain regions to better provide for their needs, and created a pay scale and a military hierarchy for what was, in effect, the first truly American coast guard (albeit at the colony/state level).[51] The following daily rations were assigned to the coast guard units: one pound of wheat bread or one and a quarter pounds of "other Bread"; one pound of pork or one and a quarter pounds of beef; "and where Saltfish can be had, they shall be allowed one Pound of Fish one Day in seven, instead of one Day's Allowance of Meat."[52] In this way, the Provincial Congress hoped to mitigate the costs associated with establishing a coast guard.

Fish also played an important role in supplying the dietary needs of privateers during the Revolution. In 1779, Richard and Knott Pedrick, Marblehead fish merchants, jointly owned, outfitted, and manned the brig *General Glover*.[53] The vessel was provisioned in part with fifty "fresh fish." In addition, it was equipped with one "cod line" valued at £2.8.0. According to the "Bill of Fair" for the Salem privateer ship *Porus* in 1781, men ate beef three days a week; pork two days; and "saltfish" every Saturday.[54] Just as was the case with fish in general, cod lines helped improve the bottom-line of fighting the war and enabled merchants and political leaders to allocate funds elsewhere.

[49] *JEPCM*, 412–413.

[50] *JHRM*, Vol. 51, Part II, 87–91, 96, 112. On April 9, 1776, the Provincial Congress raised another eight sea-coast defense companies, 50–75 men each, 400–600 men total, to guard Boston. *JHRM*, Vol. 51, Part III, 100.

[51] The British North American mainland colonies maintained temporary coast guard units as early as the 1740s. Carl E. Swanson, *Predators and Prizes: American Privateering and Imperial Warfare, 1739–1748* (University of South Carolina Press, 1991), 50–51.

[52] *JHRM*, Vol. 51, Part II, 88. The men were also given "one Shilling and eight Pence per Week to purchase Milk, Butter, Pease, or other Sauce, Soap, and Vinegar."

[53] Richard Pedrick Account Book, 1767–1784, MDHS. John Glover had been promoted to Brigadier General in 1777. George Athan Billias, *General John Glover and His Marblehead Mariners* (New York: Henry Holt and Company, 1960), 131.

[54] Logbook of the *Porus*, 1781, JDPL.

In sum, fish was heavily regulated during the Revolution, and it was used to feed American fighting men. Colonial leaders took measures to ensure provisions in general, and dried cod in particular, did not get into the hands of their enemies. Politicians within the Massachusetts Provincial Congress and the Continental Congress did not want supplies reaching either West Country migratory fishing vessels at Newfoundland or the British military forces stationed around New England. In addition, American leaders deemed it fiscally prudent to provision troops with dried, salted cod.

Though Joseph Plumb Martin may have considered dried, salted fish to be hard fare, soldiers required fuel to fight. Without fish, the American war effort would have been more expensive, and pecuniary concerns might have trumped patriotic devotion to the cause of independence. Fishermen represented another necessary ingredient for the American war effort. They are the subject of the following chapter.

11

From Fishermen to Fighting Men

> The greatest encouragement is given to seamen, which ought to be made known throughout Europe. Their pay in our Navy is eight dollars per month, with the best chance for prize money that men ever had, and liberty of discharges after every cruise, if they choose it.[1]

Parliament's Restraining Act made commercial fishing on the Atlantic Ocean illegal for colonial fishermen such as skipper Joshua Burnham and the crew of the schooner *Polly*. The maritime laborers from Ipswich, Massachusetts, did not sit idle, however. They were able to find alternate means of employment. They enlisted with the Continental Navy shortly after Congress signed the Declaration of Independence. The navy offered jobs, wages, and the promise of prize shares, which held out hope to poorer fishing families. Three of the *Polly's* crew joined Burnham on December 7, 1776, in agreeing to "ship ourselves" and "Follow all the regulations of the American Congress & be under such regulations as is Customary for Seamen & Mariners."[2] These men were not the first, nor were they the last, to make the transition from commercial fishing to military service.

[1] Robert Morris, Philadelphia merchant and member of the Continental Congress, "The present state of our navy," (1776), *AA*, Series 5, Vol. 3, 1335–1336.

[2] Joshua Burnham Papers, 1758–1817, Schooner *Polly*, 1771–1776, box 1, folder 4, JDPL. The *Polly* maintained a five-man crew during the Revolution. Joshua and Samuel Burnham, Isaac Law, and Daniel Andrews had worked on the *Polly* as fishermen before the war. Seth Story, who signed on as the schooner's cooper, was the only non-fisherman on board during the conflict. Joshua Burnham Papers, 1758–1817, Schooner *Polly*, 1771–1776, box 1, folder 4, JDPL.

In addition to fish, fishing vessels, and overseas commercial connections, American fishermen were mobilized for war. The British government's efforts to control colonists' use of the sea disaffected laborers formerly employed in the chief engine of maritime commerce in New England. Fishermen then flooded the decks of America's fighting vessels and filled the ranks of the Continental Army. They fought at sea and on land in every capacity, and their military service was invaluable to the American war effort.

The military mobilization of fishermen does not represent a uniquely American way of war any more than the conversion of fishing vessels was a late-eighteenth century invention. Various regions and peoples around the early modern Atlantic world considered the commercial fishing industry as a training school, or "nursery," for fighting men. Dutch maritime dominance had been established, according to early seventeenth-century English observers, on the foundation of its fisheries.[3] Europeans believed that fishing fleets provided vessels that could be used in fishing, trading, and warring, along with the manpower and expertise necessary for these activities. The French were then able to emulate the Dutch model through their early domination of the commercial fisheries around Newfoundland. "The whole increase of the naval greatness of France had its foundation from this trade [i.e. commercial fishing]," an English observer commented in 1745. He continued: "The French by this trade had so far increased their riches and naval power at that time [i.e. the turn of the eighteenth century], as to make all Europe stand in fear of them."[4] For its part, England waged war against the Dutch, built up its fishing industry and navy, fought against the French, and slowly engrossed the fisheries at Newfoundland. By the mid-eighteenth century, the author of *The British Merchant* could proclaim: "The history both of France and England will show you that it is since their procuring leave to fish at Newfoundland that they have grown so formidable at sea; that their navy royal has augmented

[3] See William Petyt, *Britannia Languens* (London: 1680); William de Britaine, *The Dutch Usurpation* (London: 1672); Gerard Malynes, *Lex Mercatoria* (London: 1622); and Tobias Gentleman, *England's way to win wealth, and to employ ships and mariners*... (London: 1614). For more on the economic success the Dutch experienced, see Jan De Vries and Ad Van Der Woude, *The First Modern Economy: Success, Failure, and Perseverance of the Dutch Economy, 1500–1815* (Cambridge University Press, 1997).

[4] Anonymous, *Considerations on the Trade to Newfoundland* (London: 1745). A contemporary writer referred to the French as "our most prejudicial rivals in the fishery of those parts [i.e. Newfoundland]." Anonymous, *A Short Answer to an Elaborate Pamphlet* (London: 1731).

in proportion to the numbers of ships employed in that fishery."[5] In the late seventeenth century, Sir Roger L'Estrange remarked, "The only (and the Common) Nursery of Seamen is the Fishery.... And it is well enough observed, that all Princes and States, are stronger or Weaker, at Sea, according to the Measures of their Fishery."[6] And in 1722, William Wood similarly stated: "It is a certain maxim that all states are powerful at sea as they flourish in the fishing trade."[7] Early modern English observers, in short, commonly linked manpower in the fisheries to sea power.

During the eighteenth century, British North American colonists also made the correlation between the size of the commercial fishing industry and naval supremacy. The number of men and vessels involved in the New England cod fishing industry had temporarily declined over the course of the Seven Years' War as a direct result of the use of these resources in the Royal Navy.[8] This declension led "Montesquieu" to query in a Boston newspaper in 1763: "What is become of this nursery of sailors, by which [Great Britain] have been of late aggrandized, and rendered the arbitress of the world?"[9] Then, to convince Parliamentarians not to pass legislation in 1764 and 1767 that they felt would discourage the fishing industry, members of the Boston-based Society for the Encouragement of Trade and Commerce reminded MPs on both occasions that "This Valuable Branch of our Trade," our "Fishery," represented a "Nursery for Seamen."[10] The Society argued:

> the North American Cod & Whale Fishery is a capital Article not only with respect to this Province, as it is their largest Fund for Remittances to Great Britain in payment for British Manufactures, but its National Importance is conspicuous, not only as by means of it Great Britain is

[5] *The British Merchant*, 2nd ed. (London, 1743), II, 257. Cited in Harold Adams Innis, *The Codfisheries: The History of an International Economy* (Yale University Press, 1940), 174, footnote 90.

[6] Sir Roger L'Estrange, *A discourse of the fishery briefly laying open, not only the advantages, and facility of the undertaking, but likewise the absolute necessity of it, in order to the well-being, both of king, and people*... (London: 1674).

[7] William Wood, *A Survey of Trade* (London: 1722).

[8] Innis, *The Codfisheries*, 161.

[9] *Boston Evening Post*, November 21, 1763.

[10] "In the Preamble to a late Act of Parliament," 1764; and "State of the Trade & Observations on the Late Revenue Acts," 1767, Ezekiel Price Papers, 1754–1785, MHS. The Society sent the following words to Jasper Mauduit, Massachusetts's Colonial Agent in London in 1763: "Let it be further Considered that if the Fishery here and at Newfoundland should fail, Great Britain will be deprived of a nursery for Seamen, and in a few years will want hands to Navigate her fleets. At the same time the French will have a fine opportunity to Increase their Fishery, to promote the growth of their Colonies, and put their Navy upon a respectable Footing." Thomas Cushing to Jasper Mauduit, Boston, October 28, 1763, August to December 1763 file, Jasper Mauduit Papers, 1760–1767, MHS.

> furnished with those Remittances, but also by its employing annually so great a number of vessels it constitutes a respectable Nursery of Seamen for the Navy, a principal Bulwark of the British Nation.[11]

Of course, for West Country fish merchants and sympathetic Parliamentarians the migratory fishing industry represented the only authentic nursery of seamen because of the 1699 Act and the Sixth of Anne exemption.[12] But, this shared understanding between men in England did not stop colonists from attempting to extend the claim to their industry.

Towards the end of the American Revolution, on December 11, 1781, colonial fish merchants further promulgated the popular transatlantic belief that military strength was tied to commercial fisheries. The merchants wrote to their representatives in the Massachusetts government to ensure that the "common right" of fishing in the Atlantic Ocean be "secured to the United States, whenever a Treaty of Peace shall be concluded." In particular, the merchants desired "the approaching sessions of the legislature of this Commonwealth, to move for, and to use your influence to procure, an application to Congress that they would give positive instructions to their Commissioners for negotiating a peace to make the right of the United States to the Fishery *an indispensable article of treaty*." They justified this request by stating that states'

> future rank among the Nations of the Earth will depend on their Naval Strength; and if they mean to be a commercial people, [then] it behooves them to be able at all times to protect their commerce. The means by which they can procure that protection and naval strength is to give encouragement to that kind of trade among themselves which will best serve as a nursery for seamen. The importance of the Fishery in this view is obvious from the valuable acquisitions made in the beginning of the war by our privateers, seven-eighths of which were manned from this source.[13]

[11] Petition from the Society to Parliament, sent through Dennis DeBerdt, Colonial Agent, dated 1767, Ezekiel Price Papers, 1754–1785, MHS. William Bollan, Massachusetts's Colonial Agent in London, recognized the link between the fishing industry and "national" defense. He wrote: "It need not be here observed how much this Trade of the Fishery has been the Object of the Attention of the Nation on all Occasions.... We look upon it as the chief Nursery for Seamen; and are so much interested in the other Benefits of it, that we annually send one or more of His Majesty's Ships of War [to Newfoundland], to protect our Subjects, and their Vessels, during the fishing Season." William Bollan, *The importance and advantage of Cape Breton: truly stated and impartially considered* (London: Originally printed for J. and P. Knapton, 1746; New York: Johnson Reprint Corp., 1966), 90.

[12] See the discussion in chapter six.

[13] *"Gentlemen, the inhabitants of the town of Boston..."* (Boston: Benjamin Edes and Sons, 1781), Early American Imprints, Series I: Evans #17105. Emphasis in the original.

On both sides of the Atlantic, then, throughout the seventeenth and eighteenth centuries, people viewed fishermen as vital resources in terms of military and commercial strength. That these maritime laborers played a significant role in the fighting of the Revolutionary War was, therefore, not exceptional.

Fishermen did not hesitate to fight. A case study of the military mobilization of maritime laborers in Marblehead, Massachusetts, indicates that members of this coastal community quickly made the transition from fishermen to fighting men during the conflict. Ashley Bowen, a keen resident observer of events in Marblehead, recorded in his diary at the beginning stages of military conflict on Monday, May 22, 1775, that "the fishermen are enlisting quite quick under the Congress [for the American armed forces]."[14] As mentioned in the introduction to the book, Marblehead sent more of its sons into combat than the average rural community, and the town employed more men, more vessels, and larger amounts of capital in the fish trade than any other port in the region. If fishermen were going to join the Revolution anywhere in colonial America, they would do so here.

Marblehead fishermen fought at sea in various capacities. This maritime labor included work in the coast guard, the Massachusetts Navy, the Continental Navy, and privateers. Fishermen in the coast guard were responsible for building and manning sea-coast defenses during the war.[15] These defenses consisted primarily of forts constructed at harbor entrances. Built of breastworks and field pieces, the forts guarded inbound and outbound vessels, and prevented British warships from entering coastal waters unchallenged. Permanent coast guard units were stationed in these defensive positions for the duration of the war. Marblehead, for example, maintained three "Sea Coast Defense" companies, consisting of fifty men in each company.[16]

In 1775, the men who enlisted in Marblehead's coast guard companies earned the following wages: Captains earned £5.8.0 each month; 1st Lieutenants £3.12.0; 2nd Lieutenants £3.3.0; Sergeants £2.4.0; Corporals, Fifers, and Drummers each earned £2.0.0; and Privates earned

[14] *JAB*, II, 440.

[15] Christopher P. Magra, "'Soldiers... Bred to the Sea': Maritime Marblehead, Massachusetts and the Origins and Progress of the American Revolution," *New England Quarterly*, Vol. 78, No. 4 (December 2004), 554.

[16] *MSSRW*, Vol. 5, 605, 641; Vol. 8, 230. For the size of these companies in 1775 and 1776, see *JEPCM*, 412–413; and *JHRM*, Vol. 51, Part II, 87–91, 96, 112.

"$36 [roughly £0.18.0] per month."[17] A Captain on a twelve-month enlistment, then, would have earned £64.16.0, favorably comparable to average annual wage rates most commercial fishermen earned. A man who worked on fishing expeditions and trade voyages for the same fish merchant employer, a sharesman/common seaman, could earn between £38 and £68 each year for his part in producing and distributing dried cod. A sharesman/mate could have earned between £39 and £69 annually. Skippers/masters from 1750 to 1775 in Massachusetts could earn between £45 and £75 each year.[18] A lowly Private would have earned only £10.16.0 for annual wartime work in the coast guard units. This figure is far less than the peace-time earnings most commercial fishermen commanded. Yet, men serving in these units could expect to stay in their hometowns, and there was the possibility to supplement these meager earnings through prize shares.

Coast guard units maintained vessels for use in patrolling harbors and harbor mouths, which enabled men to make captures at sea and seize supplies meant for the British military. On June 12, 1776, for example, Nathan Smith reported that he and "other persons belonging to a Seacoast company, stationed on the Island of Martha's Vineyard" captured "and brought into port, the *Bedford*, a schooner, laden with provisions and other stores, for the fleet and army." Smith petitioned the Provincial Congress for a share in the prize in return "for their risk and service."[19] Apparently, a system of prize shares had not yet been codified for the coast guard units. The men in these units, however, operated in the expectation of prize money in addition to their wages.

Marblehead fishermen such as Twisden Bowden served in the coast guard. The Bowden family had been very involved in the port's fishing industry prior to the war. Twelve different Bowdens worked for local fish merchant William Knight or local fish merchants Richard and Thomas Pedrick during the 1760s and early 1770s.[20] Twisden was born March 17, 1745, and he married into the family of a prominent local fish merchant

[17] *JEPCM*, 413. The conversion rate of $1=6 pence in Massachusetts in 1775 is drawn from John Glover's Colony Ledger, MDHS, item 729½.

[18] Christopher P. Magra, "Beyond the Banks: The Integrated Wooden Working World of Eighteenth-Century Massachusetts' Cod Fisheries," *Northern Mariner/ Le Marin Du Nord*, Vol. 17, No. 1 (January 2007; actual printing January 2008), 15.

[19] *JHRM*, Vol. 52, Part I, 28.

[20] William Knight Account Book, 1767–1781, JDPL; Richard Pedrick Account book, 1767–1784, MDHS; and Thomas Pedrick Account Book, 1760–1790, MDHS. Twisden was also described as a fisherman on a probate filed for him on November 7, 1787. *MPR*.

at age twenty.[21] He worked for Knight on at least two fares in 1770 as a sharesman on board the schooner *Barnett*, Robert Knight Jr. skipper.[22] Thirty years old and married, he enlisted as a Sergeant in Captain William Hooper's Second Marblehead Sea Coast Defense Company on July 15, 1775, and served in this capacity until December 31, 1775. His family status may have influenced his decision to join a local coast guard unit as opposed to the Continental Army, which would have taken him away from town for long periods of time. As a Sergeant in the coast guard, he would have earned £2.4.0 per month or £13.4.0 for the six months. Bowden then enlisted for a second tour of duty in Hooper's company on January 4, 1776, served for another eight months and earned an additional £17.12.0. Up to this point, then, he had earned £30.16.0 for fourteen months of coast guard duty, slightly below the annual earnings he would have made while working in the fishing industry. The following year he served in Captain Edward Fettyplace's Third Marblehead Sea Coast Defense Company. At some point in 1777, while attempting to make a capture at sea and earn additional prize money, Bowden was himself captured by the Royal Navy and sent to a prison in Halifax, Nova Scotia. In December, he was released in a prisoner exchange, and he does not appear to have gone back into the military, which may have been the result of the terms of the exchange or the experience of imprisonment.[23] Such men contributed to the military defense of the coast during the war, and their service has gone largely unrecognized.

Other fishermen chose to fight in Massachusetts's navy. Eleven states had their own flotillas during the war.[24] Massachusetts began arming vessels in 1775 and formally established its own fleet in February 1776. The wages for the officers and men in the Massachusetts Navy were as follows: Captains earned £4 each month; 1st Lieutenants made £3 per month; 2nd Lieutenants and Surgeons each earned £2.10.0 per month;

[21] *EVREC*, Vols. 1–2, 59, 46. He married Sarah Orne on December 19, 1765.

[22] William Knight Account Book, 1767–1781, JDPL.

[23] *MSSRW*, Vol. 2, 320.

[24] New Jersey and Delaware were the exceptions. See Raymond G. O'Connor, *Origins of the American Navy: Sea Power in the Colonies and the New Nation* (University Press of America, 1994), 14; and Robert Greenhalgh Albion and Jennie Barnes Pope, *Sea Lanes in Wartime: The American Experience, 1775–1942* (New York: W.W. Norton and Company, Inc., 1942), 39. For an argument that both of these states did have their own navies, see Robert L. Scheina, "A Matter of Definition: A New Jersey Navy, 1777–1783," *American Neptune*, Vol. 39 (July 1979), 209–218. More work needs to be done on these state navies to determine precisely how many vessels were involved, which type of vessels, and to develop crew lists for those vessels.

Masters earned £2; Boatswains, Carpenters, Gunners, Pilots, Quartermasters, Stewards, Master-at-Arms all earned £1.10.0 each month; while "Foremast Men" made £1.4.0 per month. In addition to these wages, and for "further Encouragement to the said Officers and Seamen," crews were to receive "one Third the Proceeds of all Captures."[25] Such prize shares provided a lucrative incentive over-and-above wages that were already attractive for unemployed fishermen. While a Captain in the state navy could have earned £48 per year, common seamen only earned £14.8.0 for the same length of service. Such compensation by itself did not compare favorably to peace-time annual earnings in the fishing industry. Prize shares, however, held out the potential for extra earnings that are difficult to quantify because of their variability.

Thomas Johnson fought in this regional navy. He was born on March 30, 1755. At age thirteen, he worked for fish merchant Thomas Pedrick as a cuttail on at least one fare on board the Marblehead schooner *Vim*. Johnson earned £7.6.11 for this fare.[26] After the Massachusetts Navy was officially organized in 1776, he signed on as a seaman on board the state brig *Massachusetts* and sailed on its first cruise in 1777. Johnson continually re-enlisted and served on three cruises under two different commanders for a total of six and a half months.[27] Given the monthly rate for fore-the-mast men in Massachusetts's navy of £1.4.0, Johnson earned £7.16.0 for his time in this sort of service, plus a share of any prizes that were taken.

Out-of-work fishermen could also earn money to support their families by fighting in the Continental Navy. Berths on naval vessels held out to maritime laborers higher than average earnings. On November 28, 1775, the Continental Congress in Philadelphia established the following pay scale for the American Navy: Captains earned £9.2.0 per month; naval "Lieutenants" £6; masters £6; mates £4.10.0; boatswains £4.10.0; boatswains' first mates £2.16.0; boatswains' second mates £2.8.0; gunners £4.10.0; gunners' mates £3.4.0; surgeons £6.8.0; surgeons' mates £4.0.0; carpenters £4.10.0; carpenters' mates £3.4.0; coopers £4.10.0; Captain's clerks £4.10.0; stewards £4.0.0; chaplains £6; able seamen £2; Captain of marines £8; marine lieutenants £5.8.0; marine sergeants £2.8.0; marine Corporals £2.4.0; fifers £2.4.0; drummers £2.4.0; and

[25] *JHRM*, Vol. 51, Part II, 256–257; and *NDAR*, Vol. 3, 1157.
[26] Thomas Pedrick Account Book, 1760–1790, MDHS.
[27] *MSSRW*, Vol. 8, 964.

"Privates or marines" £2.[28] Therefore, a Captain in the American Navy could earn £109.4.0 for a twelve-month cruise, while Privates and marines could earn £24. Put another way, Captains earned 1.5 to 2.4 times more on an annual basis than skippers/masters in the commercial fishing industry. Such wartime inflation of maritime wages was typical throughout the eighteenth-century Atlantic world in those labor markets in which naval authorities and merchants competed for manpower.[29] By contrast, privates and marines earned 1.6 to 2.8 times less than sharesman/common seaman had earned for yearly work in the commercial fishing industry. It must be remembered, however, that naval seamen were eligible for prize money. It is possible to quantify these prize shares for the American Navy.

To promote the "Activity, and Courage" of her crew, George Washington spelled out the distribution of prize money the crew of the Marblehead schooner *Hannah* would receive "over and above your Pay."[30] First, prizes were to be sent "to the nearest and safest Port," and the commander-in-chief was to be informed "immediately of such Capture, with all Particulars and there to way my further Direction." As of yet, no Admiralty courts had been established for the adjudication of legal prizes.[31] In effect, Washington told the *Hannah's* crew that he would be their ultimate judge. In the event that a prize was judged legitimate, meaning that it was neither a re-capture nor a vessel belonging to a patriot merchant, the crew would earn one-third of the value of whatever cargo was taken on a prize ship, except "military and naval stores." These stores, along with the "vessels and apparel" were "reserved for public service." In addition, the Continental Congress, which had paid for the

[28] "Rules for the Regulation of the Navy of the United Colonies," Philadelphia, November 28, 1775, in *NDAR*, Vol. 2, 1178. Again, the currency conversion ($1 = 6 shillings) is drawn from John Glover's Colony Ledger, MDHS, item 729½.

[29] Marcus Rediker, *Between the Devil and the Deep Blue Sea* (Cambridge University Press, 1987), 121–124.

[30] "George Washington's Instructions to Captain Nicholson Broughton," September 2, 1775, *NDAR*, Vol. 1, 1288. For further discussion of the *Hannah's* military conversion, see chapter nine.

[31] Thomas Cushing, a Boston merchant and Massachusetts delegate to the Continental Congress, wrote to a friend on October 23, 1775: "I am glad to find General Washington is fitting out some vessels of war. This is a necessary measure, as our enemies are daily pirating our vessels. I have frequently urged it here. As to the establishment of Courts of Admiralty, that will come on of course; but it will not do to urge it here at present." Thomas Cushing to William Cooper, Philadelphia, October 23, 1775, *Massachusetts Historical Society Collections*, Fourth Series, Vol. IV (Boston: Little, Brown, and Company, 1858), 365.

conversion of the fishing schooner, and Washington, as their commander-in-chief, received two-thirds of the cargo's remaining value. Washington then divided the crew's one-third share in the following manner: the Captain earned six shares of the one-third; the 1st Lieutenant earned five shares; the 2nd Lieutenant four; the "Ship's Master" 3; the steward 2; the mate 1.5; the gunner 1.5; the boatswain 1.5; the gunner's mate and the Sergeant each earned 1.5; and the Privates earned a singe share.[32]

Such prize shares could amount to a princely sum for unemployed workers. In 1777, the armed schooner *Franklin* and its crew captured the powder ship *Hope*. The *Hope's* precious cargo was valued at £54,075.17.2, one third of which amounted to £18,025.5.8. A single lowly Private's share for this prize, taken from the one third, was worth £487.3.4. This figure, which represents the low end of the prize shares, would have been equivalent to six-and-a-half years of work in the commercial fishing industry for even the most experienced, highest paid skippers/masters.[33] In this way, the potential for large sums of money might be offered as an incentive for unemployed, poorer working men to go to sea to capture British military supply vessels and smaller warships.

Marblehead fishermen such as Richard Tutt Jr. signed on for cruises in the American Navy during the Revolution. Tutt was the son of a fisherman. He was born on February 11, 1759, and while records of his fishing exploits have not survived, he is listed in probate records as having lived his life as a fisherman.[34] Tutt enlisted in the Marblehead regiment at the start of the Revolution and fought on land until March 20, 1776. At some point after that, he signed on as seaman on board the "U.S." brigantine *General Gates*, John Skimmer, Captain. His name appears on a list of men entitled to prize shares. Tutt was promised 1.25 shares of the crew's one-third share of any prizes taken at sea in addition to his monthly wage of £2.[35]

[32] "George Washington's Instructions to Captain Nicholson Broughton," September 2, 1775, *NDAR*, Vol. 1, 288.

[33] The American government eventually only provided the crewmen with half their legitimate shares, or £243.11.8. William Bell Clark, *George Washington's Navy: Being an Account of His Excellency's Fleet in New England* (Louisiana State University Press, 1960), 221.

[34] *EVREC*, Vol. 1, 527; and *MPR*.

[35] *MSSRW*, Vol. 16, 201. For the monthly wage of seamen in the American Navy at the end of 1775, see the wage rates listed above. It should be noted here that John Adams recommended Michael Corbet (the Marblehead mariner he had successfully defended during the *Pitt Packet* affair in 1769) for a Captain's commission in the Continental Navy. *JAB*, I, 208–209, footnote 4. It is not known whether Corbet accepted this commission

Fishermen also engaged in privateering during the American Revolution. William Le Craw was born into a (French, most likely of Jersey Island) fishing family in Marblehead on May 26, 1736.[36] Two other Le Craws can be found listed in probate records as having worked as fishermen in Marblehead, including Phillip Le Craw, who worked as a sharesman and skipper for William Knight on board the schooner *Molly* in the 1760s.[37] Like Phillip, William probably became a skipper before the war's outbreak. Unfortunately, the records of William's career in the fishing industry beyond his probate record have not survived. However, he commanded two Marblehead privateers during the war: the schooner *Necessity* (1776) and the brig *Black Snake* (1777). Fish merchants and vessel owners such as John Selman and Joshua Orne, who owned *Necessity*, would not have trusted their property and their enterprise to someone with little experience.[38] As captain of these privateers, Le Craw would have earned the highest share among the crew of any prizes taken, but no wages.[39]

While it might be expected that fishermen would fight at sea, it is perhaps less obvious that such maritime laborers would also fight on land. Yet, Marblehead fishermen who fought in the war participated on one occasion or another in some military service on *terra firma*. The local militia regiments that formed at the start of the conflict provided the first means by which Massachusetts's fishermen could supplement or replace the earnings they had lost as a result of the Restraining Act. Such local regiments then became part of the first American Army once Washington

and became an officer in the American Navy. Revolutionary service records exist for only two of the Marblehead mariners involved in the *Pitt Packet* affair. There are eight records for a "John Ryan" having done military service in the American Revolutionary War. Two of these records describe the service of men who enlisted in Marblehead. Also, there are four records for a "William Conner" having done military service in the American Revolutionary War. One record describes the service of a man who enlisted in the nearby Essex County port, Newburyport. A second record describes the man as a "Marine." For Conner, see *MSSRW*, Vol. 3, 902. For Ryan, see *MSSRW*, Vol. 13, 715.

36 He died September 20, 1802, at age sixty-six years, three months, twenty-five days. He was listed as "Captain." A probate was filed for him on October 13, 1802. *EVREC*, Vol. 2, 601; and *MPR*. It was not uncommon for French fishermen from the island of Jersey to settle in Marblehead in the eighteenth century. Christine Leigh Heyrman, *Commerce and Culture: The Maritime Communities of Colonial Massachusetts, 1690–1750* (New York: W.W. Norton & Company, 1984), 214, 246.

37 William Knight Account Book, 1767–1781, JDPL; and *MPR*.

38 *MSSRW*, Vol. 9, 624; and *JAB*, II, 664–665.

39 For first-hand accounts of life on board a privateer during the Revolution, see Sidney G. Morse, "The Yankee Privateersman of 1776," *NEQ*, Vol. 17, No. 1 (March 1944), 71–86.

assumed command. These men earned wages on top of bounties that they qualified for through their enlistment in the Continental Army.[40] In 1776, the Massachusetts Provincial Congress listed the following monthly pay rates: Colonels earned £12; Lieutenant Colonels £9.12.0; Majors £8; Captains £6; 1st Lieutenants £4; 2nd Lieutenants £3.10.0; Sergeants £2.8.0; Corporals and drummers £2.4.0; fifers and Privates £2.[41] This meant that Colonels earned £144 a year in wages, which represented 1.9 times more annual earnings than the highest earner in the fishing industry could hope for. Army Privates earned significantly less, £24 per year, which represented 1.6 times less than the lowest paid sharesman in the fishing industry. However, such enlisted men also received the aforementioned equipment bonuses and bounties as recruiting incentives. Some members of the Marblehead regiment left the ranks to board Washington's schooner fleet at the end of 1775, but others re-enlisted in the regiment when the commander-in-chief re-organized the Continental Army in January 1776.

Joseph Courtis fought on land during the Revolutionary War. He was born on October 31, 1756.[42] At age seventeen, he worked for William Knight as a cuttail on board the schooner *Molly*, Robert Knight Jr., skipper, for four fares.[43] Two years later, he enlisted as a Private in Captain Joel Smith's First Company of the 21st Massachusetts Regiment of Foot in the Continental Army under the command of Colonel John Glover on July 24, 1775. His name appears on a muster roll for August and on a company return for October. Courtis was then in the Continental

[40] It was common in 1775 for men to receive "a bounty coat or its equivalent in money" for enlisting in the Continental Army. The Continental Congress experimented for a short time in 1776 with eliminating recruiting bounties. However, the resulting lack of interest in military service soon brought the bounties back. Charles Royster, *A Revolutionary People at War: The Continental Army and American Character, 1775–1783* (University of North Carolina Press, 1979), 64–65.

[41] *JHRM*, Vol. 41, part 3, 99. This pay system further provided enlisted men with pay "at the Rate of one Day for every twenty Miles Travel" in the event that they had to walk home when their enlistment was over. And each "Non-commissioned Officer and private Soldier" were to be given "a good effective Fire-Arm and Bayonet, a Cartridge-Box, Knapsack and Blanket; and no Non-commissioned Officer and private Soldier, shall be allowed to pass Muster, without he is so equipped and provided." Ibid., 100. Connecticut's Assembly established the same wages and bounties for Privates. Harold E. Selesky, *War and Society in Colonial Connecticut* (Yale University Press, 1990), 231. According to Selesky, the same wage rates were then carried over into the Continental Army. Ibid., 235–236.

[42] *EVREC*, Vol. 1, 128.

[43] William Knight Account Book, 1767–1781, JDPL.

Army camp at Cambridge, Massachusetts, on December 20, when he applied for "a bounty coat or its equivalent in money."[44] Working as a Private in the army he earned £2 per month – £10 over five months in addition to his bounty.[45]

Marblehead fishermen further fought on both land and sea. William Main, for example, was born on October 12, 1740.[46] At twenty-nine years of age he worked on an unspecified number of fares for fish merchant Richard Pedrick.[47] On February 20, 1776, Main enlisted as a matross, a gunner's assistant who aided in loading, firing, sponging, and moving the guns in Captain Fettyplace's sea-coast defense company. Given his title, it is most probable that Main spent most of his time in the fort guarding Marblehead's harbor. He served six months and ten days "in defense of the seacoast." Assuming the pay for a matross in Massachusetts's coast guard in 1776 was equivalent to that of a private in the same type of unit in 1775 (£0.18.0 per month), Main earned roughly £5.14.0. Such money does not seem to have impressed the fisherman, and it is possible he was eager to see something other than the inside of Marblehead's fort. Eleven days before his tour of duty expired, Main signed on as a Private (most likely in a marine detachment, as his rank was not that of seaman) on board the brigantine *Massachusetts*, Daniel Souther, Captain, in the Massachusetts Navy. He served in this capacity, hoping for a chance at prize money, from August 19 to December 21, 1776. Assuming he earned the same as a fore-the-mast man, £1.4.0 per month, Main made roughly £4.16.0.[48] In total, Main most likely earned £10.10.0 for almost a year fighting on land and sea. The fact that Main and other Marblehead fishermen had experienced fighting on land and sea undoubtedly explains why such maritime laborers were among the first marines in American history.[49]

44 *MSSRW*, Vol. 4, 265.

45 This pay rate is based on the wages the Provincial Congress established in Massachusetts on April 9, 1776. *JHRM*, Vol. 51, Part III, 99.

46 He died on January 29, 1816, "in an Advanced Age." *EVREC*, Vol. 1, 330; Vol. 2, 608. A probate was filed for him on October 1, 1816, and he is listed there as a fisherman. *MPR*.

47 Richard Pedrick Account book, 1767–1784, MDHS.

48 *MSSRW*, Vol. 10, 142, 143.

49 Charles R. Smith, *Marines in the Revolution: A History of the Continental Marines in the American Revolution, 1775–1783* (Washington, DC: History and Museums Division, Headquarters, U.S. Marine Corps, 1975), 12, 32–33, 80. The Continental Congress formally established a national Marine Corps on November 10, 1775. Ibid., 7–10.

The Marblehead regiment played leading roles in some of the most famous events of the war.[50] Marbleheaders served in a supporting role at the Battle of Bunker Hill. They fought on Long Island and at Pell's Point during the White Plain's retreat. They fought at Trenton. It was their reconnaissance work that led to the capture of British General John Burgoyne at Saratoga. Later, they fought to retake Rhode Island from the British. Moreover, Marblehead fishermen were responsible for certain inland maritime activities, such as evacuating Washington and the Continental Army from Long Island and transporting those same land forces across the Delaware River prior to the Battle of Trenton.

Not every Marblehead fisherman participated in the Revolution. There are fishermen for whom there exists no record of military service. John Blackler Jr. and Thomas Foot, for example, are both listed in probate records as being fishermen.[51] Yet, neither man seems to have served in any capacity during the Revolution. Age was a factor in the decision to fight or not. Marbleheaders who performed military service were typically under the age of thirty.[52] Significantly, cod fishermen were the most physically productive in catching fish and thereby reached their peak earning potential between the ages of twenty-five and thirty.[53] In other words, those Marblehead fishermen who decided to fight against British authority lost more as a result of the Restraining Act. Those over the age of thirty, by contrast, were usually realizing fewer and fewer profits from the fishing industry. Foot was thirty-seven years old in 1775 at the start of the war, while Blackler was sixty.[54] There were also minors whose parents or legal guardians may have prevented them from serving. Sixteen was the standard age young lads were allowed into militias, although necessity ensured that there were boys under sixteen in the armed forces

[50] For more on the Marblehead regiment's military accomplishments, see George Athan Billias, *General John Glover and His Marblehead Mariners* (New York: Henry Holt and Company, 1960).

[51] Probates were filed for Blackler and Foot on July 11, 1787, and December 8, 1785, respectively. *MPR*.

[52] Walter Leslie Sargent, "Answering the Call to Arms: The Social Composition of the Revolutionary Soldiers of Massachusetts, 1775–1783." (Ph.D. Dissertation, University of Minnesota, 2004), 228, table 6.8, 229, figure 6.9; and William Arthur Baller, "Military Mobilization during the American Revolution in Marblehead and Worcester, Massachusetts" (Ph.D. Dissertation, Clark University, 1994), 27, 366, figure 4, and 367, figure 5.

[53] Daniel Vickers, *Farmers and Fishermen: Two Centuries of Work in Essex County, Massachusetts, 1630–1830* (University of North Carolina Press, 1994), 178–180, esp. figure 2.

[54] *EVREC*, Vol. 1, 45; and Vol. 1, 179.

during the war.[55] Regardless of their motivations for not fighting for American independence, Loyalists and neutrals had to be very careful in fishing communities. Patriots persecuted those who publicly supported the Crown and Parliament and rode them out of Marblehead very early in 1775.[56]

There is no evidence of African-American fishermen having served in the war, but then again there is very little evidence of African Americans working in the cod fishing industry.[57] There *is* strong evidence that African Americans served in the Marblehead regiment, however.[58] Cato Prince, described as "a Blackman belonging to said Marblehead," had been a slave owned by the Prince family in the fishing community. Cato was imported from Africa and purchased sometime between 1765 and 1770 by Captain John Prince. Around 1770, however, Captain Prince freed Cato from bondage. Then, in 1780, Cato "went as a Free-man into the Service of the United States as a Soldier." In his 1818 pension deposition, Prince stated, "When I first enlisted as a Soldier, I enlisted for 6 months; but when the 6 months were up, I was turned over for a three year Man." In this manner, Prince came into the war late but maintained his service until the end. He entered the war as a private and left the war with the same rank. For his entire service, he received "4 LB Sugar, a Jack knife, one Coat, Waistcoat and Britches, 4 pair of Stockings, 4 Shirts, 1 pair Gaiters [protective covering overlapping lower part of britches and shoes] and 1 Handkerchief." On April 18, 1818, a sixty-two-year-old Cato Prince received a pension from the U.S. government of eight dollars a month for his military service.[59]

Additionally, Alexander Graydon, an officer in the Continental Army from Pennsylvania, observed and described the forces from Massachusetts early in 1776. Graydon derisively singled out the Marblehead unit, noting

[55] Don Higginbotham, *The War of American Independence: Military Attitudes, Policies, and Practice, 1763–1789*, 2nd ed., (Northeastern University Press, 1983), 391.

[56] See the first-hand accounts described in "Journal of Rev. Joshua Wingate Weeks, Loyalist Rector of St. Michael's Church, Marblehead, 1778–1779," *Essex Institute Historical Collections*, Vol. 52 (Salem, MA: Essex Institute Press, 1916), 1–16, 161–176, 197–208, 345–356; and "Essex County Loyalists," *Essex Institute Historical Collections*, Vol. 43 (Salem, MA: Essex Institute Press, 1907), 289–316.

[57] See the discussion in chapter three. The 1771 Tax List reveals eighteen slaves, listed as "Servants For Life," in Marblehead at this time. Bettye Hobbs Pruitt, ed., *The Massachusetts Tax Valuation List of 1771* (Boston: G. K. Hall & Co., 1978), 98–107.

[58] Magra, "'Soldiers . . . Bred to the Sea'," 549–550.

[59] *Pension Records*, MDHS, item 7578. Secretary of War, John C. Calhoun, approved the pension of this African-American patriot.

that "in this regiment there were a number of Negroes, which to persons unaccustomed to such associations, had a disagreeable, degrading effect."[60] African-American service in this fishing community's regiment was most likely a function of fish merchants' commercial ties to the Southern and West Indian plantations. Slaves could have been purchased, as they were in Rhode Island, and imported as chattel.[61] Just as the slaves in the Southern plantations fled to Lord Dunmore after his emancipation proclamation, slaves in New England probably decided to fight for whichever side they believed offered them the best chance at a life in freedom.[62]

Since there was no indication of women having worked in the industry prior to the war, it comes as no surprise that there is no evidence of female workers in the fishing industry in Marblehead providing military service during the Revolution.[63] Marblehead's military regiments mirrored the society that created it. In general, women certainly played supporting roles as camp followers for the Continental Army, and a few females masqueraded as men and actually fought in the Army.[64] Yet, the same gender norms that kept women on the margins of the commercial cod fishing industry in New England seem to have shaped the contours of their military service and relegated them to the home front in fishing ports.

The British almost certainly used fishermen in their military. The ebb and flow of the war effort for the British took on some of the characteristics of the migratory fishing industry in peace time. Whereas West Country migratory vessels and their naval convoys typically picked up laborers in Ireland before heading to Newfoundland, the British navy and army stopped in Irish ports and recruited thousands of Irish men for

[60] Quoted in Benjamin Quarles, *The Negro in the American Revolution* (University of North Carolina Press, 1961), 72.

[61] See Robert K. Fitts, *Inventing New England's Slave Paradise: Master/Slave Relations in Eighteenth-Century Narragansett, Rhode Island* (New York: Garland Pub., 1998).

[62] See Cassandra Pybus, *Epic Journeys of Freedom: Runaway Slaves of the American Revolution and Their Global Quest for Liberty* (Boston: Beacon Press, 2006); Michael Lee Lanning, *Defenders of Liberty: African Americans in the Revolutionary War* (New York: Citadel Press, 2000); Sylvia R. Frey, *Water from the Rock: Black Resistance in a Revolutionary Age* (Princeton University Press, 1991); and Quarles, *The Negro in the American Revolution.*

[63] See the discussion in chapter three.

[64] See Holly A. Mayer, *Belonging to the Army: Camp Followers and Community during the American Revolution* (University of South Carolina Press, 1996); and Alfred Young, *Masquerade: The Life and Times of Deborah Sampson, Continental Soldier* (New York: Alfred A. Knopf, 2004).

military duty before heading to Boston in 1775.[65] It is very likely that at least some of these men were among the same maritime laborers who had gone westward to Newfoundland before the Revolution.[66]

In sum, fishermen were mobilized in a wide variety of ways for the American Revolutionary War. Fishermen fought at sea on state and Continental naval vessels, in the coast guard, and on privateers. These maritime laborers also fought on land in the Continental Army. America's armed forces offered these out-of-work maritime laborers the chance to strike a blow at a government that had threatened their livelihoods. In addition, the American military provided employment, wartime wages, and certain financial bonuses that helped support fishing families while their industry was closed. Without coming to terms with the nature of the work and the maritime laborers in the cod fisheries prior to the Revolution, it is difficult to understand who fought and why.

Fishermen were invaluable to American military commanders during the Revolutionary War. On October 13, 1775, George Washington wrote to his brother, John Augustine Washington, that fishermen were "soldiers . . . who had been bred to the sea."[67] One year later, following the evacuation of Long Island, the commander-in-chief informed the Continental Congress: "I must depend upon them [i.e. fishermen] for a successful opposition to the Enemy."[68] In the middle of 1776 fishermen saved Washington and the Continental Army from British capture on Long Island by successfully ferrying 9,000 American troops to safety under the cover of fog.[69] It was also fishermen who were responsible for transporting Washington and the Continental Army safely across the Delaware River at the end of 1776, allowing the victory at Trenton and

[65] For more on the employment of Irish fishermen in the migratory fishing industry at Newfoundland in the eighteenth century, see W. Gordon Handcock, *Soe longe as there comes noe women: Origins of English Settlement in Newfoundland* (St. Johns, NL: Breakwater Books, 1989), 30–31, 133–134. For evidence that British naval vessels stopped at Irish ports for manpower before coming to America, see *NDAR*, Vol. 1, 333–336.

[66] More work needs to be done on the role fishermen and the fishing industry played in the British armed forces during the Revolution. Current research on the British military has not explored the role the commercial fishing industry played in providing manpower and vessels for the war effort.

[67] John C. Fitzpatrick, ed., *The Writings of George Washington, 1745–1799*, Vol. 4 (Washington, DC: U.S. Government Printing Office, 1934), 27.

[68] Ibid., Vol. V, 501.

[69] For a full account, see Billias, *General John Glover and His Marblehead Mariners*, 96–104.

the capture of much needed supplies.[70] General Henry Knox recalled Washington's crossing of the Delaware River later during a meeting of the Massachusetts legislature. He stated:

> I wish the members of this body could have stood on the banks of the Delaware River in 1776, in that bitter night when the commander-in-chief had drawn up his little army to cross it, and had seen the powerful current beating onward the floating masses of ice which threatened destruction to whoever should venture upon it. I wish that when this occurrence threatened to defeat the enterprise, that they could have heard that distinguished warrior demand, 'Who will lead us on?' and seen the men from Marblehead, and Marblehead alone, stand forth to lead the army along the perilous path to Trenton. There, sirs, went fishermen of Marblehead, alike at home upon land or water.[71]

Impassioned rhetoric aside, Knox's first-hand insight into the occupational identity of the men responsible for the hazardous winter crossing should not be overlooked or underestimated. Fishermen participated in a variety of ways in the Revolutionary War, and they made key contributions to the war effort that secured American independence.

[70] For a full account, see Ibid., 3–24. The most recent systematic treatment of the crossing only briefly mentions the fishermen. David Hackett Fischer, *Washington's Crossing* (Oxford University Press, 2004), 21–22.

[71] Quoted in "Memoir Read Before the Massachusetts Historical Society," Joseph Williamson, 16 November 1881, *Massachusetts Historical Society Collections*.

Conclusion

Commercial fishing was even connected to the peace process that ended the Revolutionary War. American leaders formally opened negotiations for peace with Great Britain in 1779. According to Edmund Cody Burnett, the foremost authority on the Continental Congress, these leaders resolved to demand six stipulations for peace over and above "absolute and unlimited . . . liberty, sovereignty, and independence."[1] "The most hotly contested parliamentary battle ever waged in Congress," Burnett writes, involved the stipulation regarding "fishing rights on the banks and coasts of Newfoundland."[2] The Grand Bank, which was the richest source of cod in the Atlantic Ocean, was one of the spoils of the war.

New England delegates stubbornly insisted that Americans wage war until access to these fishing waters was guaranteed. These rights were, after all, one of the foremost causes of the imperial conflict in the first place. The issue split Congress into two geographical factions: a Southern faction that supported peace without fishing rights, and a Northern faction, which New Englanders championed, that would not accept a treaty giving up "the common right of fishing."[3] Congress remained deadlocked, and the peace process utterly broke down until middle ground was reached on this important issue.

In the compromise, Northern delegates agreed not to make fishing rights an absolute ultimatum for peace, and the Southern members of

[1] Edmund Cody Burnett, *The Continental Congress* (New York: MacMillan Company, 1941), 431.

[2] Ibid., 431, 433.

[3] Ibid., 434.

Congress agreed to allow a New England delegate to help negotiate peace terms with Great Britain. In the end, John Adams, along with Ben Franklin and John Jay, secured American access to fishing waters around Newfoundland, including the Grand Bank, in Article III of the Treaty of Paris in 1783.[4] From start to finish, then, commercial fishing played important roles in the American Revolution.

Investigating the ties between commercial fishing and the American Revolution reveals that the origins and progress of this formative event cannot be fully explained without putting the Atlantic Ocean at the center of the independence story. The British North American colonies were part of a wider maritime world. They relied on access to oceanic resources and overseas markets. Americans were not solely self-sufficient farmers who remained isolated from the ill-effects of imperial commercial regulations until very late in the eighteenth century. Fishermen who plied their trade on the ocean helped make the Revolution possible, and they contributed much to the American war effort. Their cause should be remembered alongside farmers.

The Atlantic Ocean factored mightily in the origins of the American Revolution. The ocean provided a key source of employment for colonial laborers. In eighteenth-century New England, where available land was dwindling and jobs were scarce, going to sea was often the only alternative for coastal peoples. Fishermen risked life and limb on offshore banks to provide for their families. Commercial fishing further stimulated work in shipbuilding and lumbering. When the British government threatened the colonial cod fisheries with harmful legislation, it was, in effect, posing a danger to a key source of jobs in a region experiencing shrinking occupational opportunities. When workers' access to oceanic resources and overseas shipping lanes was restricted, they resisted what amounted to government attempts to deny them the right to work.

A similar story held true for colonial fish merchants. The Atlantic Ocean provided profits to entrepreneurs willing and financially able to take risks. Dried, salted cod represented one of the most valuable Atlantic commodities. Producing and distributing this trade good was a profitable colonial enterprise. When the British government limited merchants' ability to extract resources from the ocean, and when the state restricted merchants' access to overseas markets, merchants resisted the state's sovereignty of the seas.

[4] Clive Parry, ed., *The Consolidated Treaty Series*, Vol. 48 (New York: Oceana Publications, Inc., 1969), 492–493.

The ocean was also the site and source of considerable intra-imperial business competition, which helped bring about the Revolution. Fish merchants on both sides of the Atlantic struggled for control over the best fishing waters and the most lucrative overseas fish markets. Entrepreneurs in England enjoyed decided advantages in this maritime rivalry. They were best able to mobilize state power to protect their commercial interests, which caused colonial resentment to grow into revolutionary fervor.

In addition to factoring into the origins of the Revolution, the Atlantic Ocean enabled American colonists to defeat the British government in war. Fish merchants turned their willingness to take risks and their risk management strategies to the way of war. The American government bent merchants' profit motive to the war effort. Transoceanic trade routes and overseas business contacts provided access to military supplies that were fundamentally necessary for the prosecution of the American war effort. Military intelligence also flowed through these supply lines, as overseas merchants informed Americans of British troop movements in Europe. Such knowledge added to the combat effectiveness of the military stores.

The ocean further provided a wide open arena for combat that enabled Americans to take the fight to the British in different ways. Rather than sit on the shoreline and watch the British supply and transport troops by sea, the Americans armed vessels and engaged the enemy. The Continental Navy, state navies, coast guard units, and privateers cut supply lines and captured prizes. These efforts provided vessels and supplies for the American war effort and weakened Britain's ability to fight.

The ocean did pose several challenges to the American war effort, however. American political leaders were very concerned about the use of oceanic resources. In particular, they did not want fish reaching migratory West Country vessels in North Atlantic waters around Newfoundland. Neither did they want provisions of any sort reaching British troops stationed in North America. At the same time, American leaders recognized that the ocean provided a comparatively inexpensive source of protein that could be used to feed their fighting men. These leaders further acknowledged that fish could be used in overseas trade in exchange for military supplies. They, therefore, regulated the overseas transport of dried, salted cod. This regulation helped Americans control the transatlantic flow of maritime resources, which in turn helped them to prosecute the war.

To a certain degree, the ocean also provided manpower for the American war effort. Fishermen, men who earned their living fighting against and learning from the deep blue sea, fought in every capacity against the British. They had especially angry axes to grind against the British government because of the Restraining Act. They joined in the Continental Army and engaged in combat on land. They enlisted in the Continental Navy, state navies, coast guard units, and they signed on to privateers, all to fight against the British at sea. Not every colonial fisherman fought in the Revolutionary War. But those who did participated in a variety of ways, and American commanders especially valued their military service. In these ways, then, the Atlantic Ocean facilitated the progress of the American war effort and factored into the birth of the United States of America.

Index